CRAN Recipes

DPLYR, Stringr, Lubridate, and RegEx in R

William Yarberry

Apress®

CRAN Recipes: DPLYR, Stringr, Lubridate, and RegEx in R

William Yarberry
Kingwood, TX, USA

ISBN-13 (pbk): 978-1-4842-6875-9   ISBN-13 (electronic): 978-1-4842-6876-6
https://doi.org/10.1007/978-1-4842-6876-6

Managing Director, Apress Media LLC: Welmoed Spahr
Acquisitions Editor: Steve Anglin
Development Editor: Matthew Moodie
Coordinating Editor: Mark Powers

Cover designed by eStudioCalamar

Cover image by Fly-D on Unsplash (www.unsplash.com)

Distributed to the book trade worldwide by Apress Media, LLC, 1 New York Plaza, New York, NY 10004, U.S.A. Phone 1-800-SPRINGER, fax (201) 348-4505, e-mail orders-ny@springer-sbm.com, or visit www.springeronline.com. Apress Media, LLC is a California LLC and the sole member (owner) is Springer Science + Business Media Finance Inc (SSBM Finance Inc). SSBM Finance Inc is a **Delaware** corporation.

For information on translations, please e-mail booktranslations@springernature.com; for reprint, paperback, or audio rights, please e-mail bookpermissions@springernature.com.

Apress titles may be purchased in bulk for academic, corporate, or promotional use. eBook versions and licenses are also available for most titles. For more information, reference our Print and eBook Bulk Sales web page at http://www.apress.com/bulk-sales.

Any source code or other supplementary material referenced by the author in this book is available to readers on GitHub via the book's product page, located at www.apress.com/9781484268759. For more detailed information, please visit http://www.apress.com/source-code.

Printed on acid-free paper

Table of Contents

About the Author

Bill Yarberry, CPA, is Principal Consultant, ICCM Consulting LLC, based in Houston, Texas. In addition to writing business and technical books, he consults in the areas of small business development, IT governance, security consulting, project management, and business analytics. He was previously a senior manager with PricewaterhouseCoopers, responsible for telecom and network services in the Southwest region. Yarberry has more than 20 years of experience in a variety of IT-related services, including application development, data analytics, internal audit management, outsourcing administration, and Sarbanes-Oxley consulting. As a writer, his focus has been on providing readers with the fastest route to subject matter competence, in both business and technical domains.

His books include *The Effective CIO* (co-authored), *What Top CIOs Know* (co-authored), *$250K Consulting*, *Computer Telephony Integration*, *Telecommunications Cost Management*, *GDPR: A Short Primer*, and *Write a Business Plan to Make Money*. In addition, he has written over 20 professional articles on topics ranging from wireless security to change management. One of his articles, "Audit Rights in an Outsource Environment," received the Institute of Internal Auditors Outstanding Contributor Award. He maintains a blog on his author page, www.amazon.com/author/billyarberry.

Prior to joining PricewaterhouseCoopers, Yarberry was Director of Telephony Services for Enron Corporation. He was responsible for operations, planning, and architectural design for voice communications servers and related systems for more than 7000 employees. Yarberry graduated Phi Beta Kappa in Chemistry from the University of Tennessee and earned an MBA at the University of Memphis. He enjoys reading history, future studies, antique pen collecting, swimming, occasional scuba diving, spotting statistical nonsense in official publications, and spending time with family.

He welcomes comments and suggestions and can be contacted at byarberry@ iccmconsulting.net.

About the Technical Reviewer

 **Jon Westfall** is an award-winning professor, author, and practicing cognitive scientist. He teaches a variety of courses in psychology, from introduction to psychology to graduate seminars. His current research focuses on the variables that influence economic and consumer finance decisions, as well as retention and persistence of college students. With applications to psychology, information technology, and marketing, his work finds an intersection between basic and applied science. His current appointments include Associate Professor of Psychology, Coordinator of the First Year Seminar Program, and Coordinator of the Psychology Program at Delta State University. Prior to joining the faculty at Delta State in 2014, he was Visiting Assistant Professor at Centenary College of Louisiana and Associate Director for Research and Technology at the Center for Decision Sciences, a center within Columbia Business School at Columbia University in New York City. He now maintains a role with Columbia Business School as a research affiliate/variable hours officer of administration and technology consultant.

Acknowledgments

Samuel Johnson, the famous 18th-century writer and creator of the first major English dictionary, said, "He who is self-taught has a blockhead for a teacher." To some extent, that applies to self-editing as well. It is rare indeed to find people who can find all their own grammatical faults. With that in mind, I give many thanks to my wife, Carol, whose artwork, editing, and acumen for discovering my grammatical flaws never cease to amaze me.

I also recognize vital people in my life over the years—my children, Will and Libby; my parents; and my high school science teacher, Mrs. Jones.

Introduction

Why yet another basic R book? Answer: to get you up and running with practical R knowledge in the shortest time possible. With this narrow purpose in mind, I have omitted extended discussions, side trips into exotic data science domains, and warnings of possible misuse of statistics. This book uses common words where the technical meaning is clear. For example, the word "origin" as used in Lubridate can be replaced by the English equivalent "starting date and time." Short explanations, along with the accompanying R code, allow you to cut and paste useful code into your applications. Early success motivates.

I've always thought that the title of so many nonfiction books, such as "X for Dummies," gets it wrong. The problem is not a limitation of smarts; it's a limitation of time. With millions of potential subject matter topics in business, science, and society, everyone must narrow their selection of deep dive areas. You need lots of tools but must limit your domain expertise to a handful of areas. For everything else, "good enough" is the sweet spot. Learning automotive engineering is overkill for simply driving a car. *CRAN Recipes* is not a 50-pound definitive book on statistics and R. It's designed to get you to your objectives quickly and accurately.

All code in this book has been tested under Windows R version 4.0.3, using RStudio version 1.3.1093 as the integrated development environment (IDE). In most cases, knitr provided the formatting shown in the book. In other environments, such as Mac or Ubuntu Linux, a few of the packages may not load automatically, due to different dependencies. These will need to be loaded separately. For example, the crayon package will not automatically load with tidyverse under Ubuntu Linux. Nonetheless, 99% of the code shown here should work unchanged in a non-Windows environment.

In some cases, different packages use the same function names. For example, the select statement for DPLYR and MASS has the same name. There are two ways to get around this problem:

- Load tidyverse (or in this case DPLYR) immediately before using the select statement if you have recently used the MASS package.

- Specify the package when executing the command. For example, "dplyr::select ....." will ensure that the DPLYR version of select is used (generally the best option).

Since this book favors tidyverse above other package systems (unless otherwise noted), you should code library(tidyverse) after using MASS or other conflicting packages. Perhaps the best solution is to specify the package associated with the command, as shown earlier.

Code to run prior to examples in this book

For executing code in this book, first execute the following. It is smart enough to look at your system, decide whether you have tidyverse already installed, install if needed, and load into memory automatically.

```
if (!require("tidyverse")) install.packages("tidyverse")
```

Packages Lubridate, DPLYR, and Stringr, among others, are built into tidyverse. Run the following code to see all packages included in tidyverse:

```
tidyverse_packages()
```

```
##  [1] "broom"       "cli"         "crayon"      "dbplyr"      "dplyr"
##  [6] "forcats"     "ggplot2"     "haven"       "hms"         "httr"
## [11] "jsonlite"    "lubridate"   "magrittr"    "modelr"      "pillar"
## [16] "purrr"       "readr"       "readxl"      "reprex"      "rlang"
## [21] "rstudioapi"  "rvest"       "stringr"     "tibble"      "tidyr"
## [26] "xml2"        "tidyverse"
```

Feel free to use this book's code for any purpose, recognizing of course that no warranty is provided. This code is for pedagogical purposes only. In some cases, I use "bk." (for "book") to start variables to clearly distinguish created variables from reserved words. It is overkill for experienced R people. However, I've seen some texts where, for example, "keep" has been comingled with a temporary variable named "keeps" or "mutate" with "Mutate" and so on. Yes, these minor changes do serve technically to distinguish them from base R or system terms. However, it is dysfunctional for teaching new users to code. In this book and code, when you see "bk.", it will always be a variable and never a command, functional specification, or otherwise reserved word.

In later chapters, I'll bring in a few specialty packages to demonstrate some interesting visualization and other capabilities of R. In those cases, package names will be included, along with the assumption that tidyverse has already been loaded.

For the most part, the examples in this book have been created by running code in RStudio, using the knitr package. Knitr sends executed code to either html, Microsoft Word, or PDF. Output lines are shown with a "##" on the left of each line. If you want to see how this works, create a simple R script in RStudio, save it, and then press (in Windows) Ctrl-Shift-K. On Mac, press Command-Shift-K. Figure 1 shows the first part of the dialog.

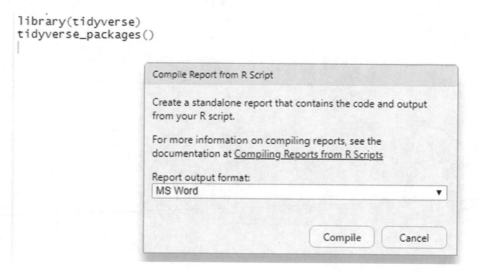

Figure 1. *First step in creating reproducible output using knitr (Ctrl-Shift-K)*

CHAPTER 1

DPLYR

DPLYR is one of my favorite R packages. Its logical and consistent rules replace the older, motley collection of syntactically inconsistent packages and functions. It's like a Swiss Army knife in the woods—don't leave home without it.

Most of the book's code examples use built-in R datasets or toy dataframe hard-coded into the program. For practice, you should substitute your own data when running the snippets of code.

1.1 Filter Commands

The filter command is used to eliminate rows (records) you do not want. The following commands use built-in datasets as the input dataframe. The dataset "mtcars" is used in the following. The output shows cars with six cylinders only.

Note The following shown libraries will be used in all code unless otherwise noted. DPLYR is included in the mega-package tidyverse.

1.1.1 Single-Condition Filter

```
library(tidyverse)
data("mtcars")
#select only cars with six cylinders
six.cyl.only <- filter(mtcars, cyl == 6)
six.cyl.only
```

© William Yarberry 2021
W. Yarberry, *CRAN Recipes*, https://doi.org/10.1007/978-1-4842-6876-6_1

```
##                      mpg cyl  disp  hp drat    wt  qsec vs am gear carb
## Mazda RX4           21.0   6 160.0 110 3.90 2.620 16.46  0  1    4    4
## Mazda RX4 Wag       21.0   6 160.0 110 3.90 2.875 17.02  0  1    4    4
## Hornet 4 Drive      21.4   6 258.0 110 3.08 3.215 19.44  1  0    3    1
## Valiant             18.1   6 225.0 105 2.76 3.460 20.22  1  0    3    1
## Merc 280            19.2   6 167.6 123 3.92 3.440 18.30  1  0    4    4
## Merc 280C           17.8   6 167.6 123 3.92 3.440 18.90  1  0    4    4
## Ferrari Dino        19.7   6 145.0 175 3.62 2.770 15.50  0  1    5    6
```

In the **filter** command, "equals" is a double equals sign "==".

1.1.2 Multiple-Condition Filter

Filter the dataset mtcars for both six cylinders and 110 horsepower:

```
six.cylinders.and.110.horse.power <- filter(mtcars, cyl == 6,
  hp == 110)
six.cylinders.and.110.horse.power
```

```
##                      mpg cyl disp  hp drat    wt  qsec vs am gear carb
## Mazda RX4           21.0   6  160 110 3.90 2.620 16.46  0  1    4    4
## Mazda RX4 Wag       21.0   6  160 110 3.90 2.875 17.02  0  1    4    4
## Hornet 4 Drive      21.4   6  258 110 3.08 3.215 19.44  1  0    3    1
```

1.1.3 OR Logic for Filtering

You can use as many OR symbols (pipe |) as needed.

Filter based on the OR logical operator:

```
gear.eq.4.or.more.than.8 <- filter(mtcars, gear == 4|cyl > 6)
gear.eq.4.or.more.than.8
```

```
##                      mpg cyl  disp  hp drat    wt  qsec vs am gear carb
## Mazda RX4           21.0   6 160.0 110 3.90 2.620 16.46  0  1    4    4
## Mazda RX4 Wag       21.0   6 160.0 110 3.90 2.875 17.02  0  1    4    4
## Datsun 710          22.8   4 108.0  93 3.85 2.320 18.61  1  1    4    1
```

```
## Hornet Sportabout    18.7   8 360.0 175 3.15 3.440 17.02  0  0    3    2
## Duster 360           14.3   8 360.0 245 3.21 3.570 15.84  0  0    3    4
## Merc 240D            24.4   4 146.7  62 3.69 3.190 20.00  1  0    4    2
## Merc 230             22.8   4 140.8  95 3.92 3.150 22.90  1  0    4    2
## Merc 280             19.2   6 167.6 123 3.92 3.440 18.30  1  0    4    4
## Merc 280C            17.8   6 167.6 123 3.92 3.440 18.90  1  0    4    4
## Merc 450SE           16.4   8 275.8 180 3.07 4.070 17.40  0  0    3    3
## Merc 450SL           17.3   8 275.8 180 3.07 3.730 17.60  0  0    3    3
## Merc 450SLC          15.2   8 275.8 180 3.07 3.780 18.00  0  0    3    3
## Cadillac Fleetwood   10.4   8 472.0 205 2.93 5.250 17.98  0  0    3    4
## Lincoln Continental  10.4   8 460.0 215 3.00 5.424 17.82  0  0    3    4
## Chrysler Imperial    14.7   8 440.0 230 3.23 5.345 17.42  0  0    3    4
## Fiat 128             32.4   4  78.7  66 4.08 2.200 19.47  1  1    4    1
## Honda Civic          30.4   4  75.7  52 4.93 1.615 18.52  1  1    4    2
## Toyota Corolla       33.9   4  71.1  65 4.22 1.835 19.90  1  1    4    1
## Dodge Challenger     15.5   8 318.0 150 2.76 3.520 16.87  0  0    3    2
## AMC Javelin          15.2   8 304.0 150 3.15 3.435 17.30  0  0    3    2
## Camaro Z28           13.3   8 350.0 245 3.73 3.840 15.41  0  0    3    4
## Pontiac Firebird     19.2   8 400.0 175 3.08 3.845 17.05  0  0    3    2
## Fiat X1-9            27.3   4  79.0  66 4.08 1.935 18.90  1  1    4    1
## Ford Pantera L       15.8   8 351.0 264 4.22 3.170 14.50  0  1    5    4
## Maserati Bora        15.0   8 301.0 335 3.54 3.570 14.60  0  1    5    8
## Volvo 142E           21.4   4 121.0 109 4.11 2.780 18.60  1  1    4    2
```

1.1.4 Filter by Minimums, Maximums, and Other Numeric Criteria

The output shows, as one would expect, a single row with the smallest engine displacement:

```
smallest.engine.displacement <- filter(mtcars, disp ==
    min(disp))
smallest.engine.displacement
```

```
##                mpg cyl disp hp drat    wt qsec vs am gear carb
## Toyota Corolla 33.9   4 71.1 65 4.22 1.835 19.9  1  1    4    1
```

Filter with conditions separated by commas:

```
data("ChickWeight")
chick.subset <- filter(ChickWeight, Time < 3, weight > 53)
chick.subset
```

```
##    weight Time Chick Diet
## 1      55    2    22    2
## 2      55    2    40    3
## 3      55    2    43    4
## 4      54    2    50    4
```

1.1.5 Filter Out Missing Values (NAs) for a Specific Column

The built-in dataset "airquality" has a missing value in the fifth row of the first column (Ozone):

```
data("airquality")
head(airquality,10) #before filter
```

```
##    Ozone Solar.R Wind Temp Month Day
## 1     41     190  7.4   67     5   1
## 2     36     118  8.0   72     5   2
## 3     12     149 12.6   74     5   3
## 4     18     313 11.5   62     5   4
## 5     NA      NA 14.3   56     5   5
## 6     28      NA 14.9   66     5   6
## 7     23     299  8.6   65     5   7
## 8     19      99 13.8   59     5   8
## 9      8      19 20.1   61     5   9
## 10    NA     194  8.6   69     5  10
```

Remove any row with missing values in the Ozone column:

```
no.missing.ozone = filter(airquality, !is.na(Ozone))
head(no.missing.ozone,8) #after filter
```

```
##   Ozone Solar.R Wind Temp Month Day
## 1    41     190  7.4   67     5   1
## 2    36     118  8.0   72     5   2
## 3    12     149 12.6   74     5   3
## 4    18     313 11.5   62     5   4
## 5    28      NA 14.9   66     5   6
## 6    23     299  8.6   65     5   7
## 7    19      99 13.8   59     5   8
## 8     8      19 20.1   61     5   9
```

Note that although the row with NA for Ozone has been eliminated, the row with an NA for Solar.R is still there.

1.1.6 Filter Rows with NAs Anywhere in the Dataset

Use complete.cases() to remove any rows containing an NA in any column:

```
airqual.no.NA.anywhere <- filter(airquality[1:10,],
  complete.cases(airquality[1:10,]))
airqual.no.NA.anywhere
```

```
##   Ozone Solar.R Wind Temp Month Day
## 1    41     190  7.4   67     5   1
## 2    36     118  8.0   72     5   2
## 3    12     149 12.6   74     5   3
## 4    18     313 11.5   62     5   4
## 5    23     299  8.6   65     5   7
## 6    19      99 13.8   59     5   8
## 7     8      19 20.1   61     5   9
```

1.1.7 Filter by %in%

"%in%" is a powerful operator, providing a convenient shorthand for including/
excluding specified values:

```
data("iris")
table(iris$Species) #counts of species in the dataset

##
##     setosa versicolor  virginica
##         50         50         50

iris.two.species <- filter(iris,
Species %in% c("setosa", "virginica"))
table(iris.two.species$Species)

##
##     setosa versicolor  virginica
##         50          0         50
```

Show the number of rows before and after filtering:

```
nrow(iris); nrow(iris.two.species)

## [1] 150
## [1] 100
```

1.1.8 Filter for Ozone > 29 and Include Only Three Columns

```
data("airquality")
airqual.3.columns <- filter(airquality, Ozone > 29)[,1:3]
head(airqual.3.columns)

##    Ozone Solar.R Wind
## 1     41     190  7.4
## 2     36     118  8.0
## 3     34     307 12.0
```

```
## 4     30      322 11.5
## 5     32       92 12.0
## 6     45      252 14.9
```

1.1.9 Filter by Total Frequency of a Value Across All Rows

This logic uses "group_by" to enable counting of rows based on number of gears. After the counts of gears are made, then only those rows whose **total** counts exceed ten are included in the output. All you want to see here are records that have at least 11 rows with a specific number of gears in the car. The filter is driven solely by frequency of occurrence. Your question may be phrased as "just show me records where common gear configurations occur." Five gears are not nearly as common as three and four, so in the filtered dataframe, they are omitted. In the following first table, there are 15 records with a car having three gears, 12 records for four gears, and five records for five gears. After applying the filter and creating a new dataframe, there are no records having five gears:

```
table(mtcars$gear)
```

```
##
##  3  4  5
## 15 12  5
```

```
more.frequent.no.of.gears <- mtcars %>%
  group_by(gear) %>%
  filter(n() > 10)  #
table(more.frequent.no.of.gears$gear)
```

```
##
##  3  4
## 15 12
```

Additional criteria can be added to the filter by including a requirement that the horsepower be less than 105:

```
more.frequent.no.of.gears.and.low.horsepower <- mtcars %>%
group_by(gear) %>%
  filter(n() > 10, hp < 105)
  table(more.frequent.no.of.gears.and.low.horsepower$gear)

##
## 3 4
## 1 7
```

1.1.10 Filter by Column Name Using "starts with"

In this code, records are selected where the column name starts with an "S":

```
names(iris)  #show the column names

## [1] "Sepal.Length" "Sepal.Width"  "Petal.Length" "Petal.
Width"  "Species"

iris.display <- iris %>% dplyr::select(starts_with("S"))
head(iris.display)  #use head to reduce number of rows output

##   Sepal.Length Sepal.Width Species
## 1          5.1         3.5 setosa
## 2          4.9         3.0 setosa
## 3          4.7         3.2 setosa
## 4          4.6         3.1 setosa
## 5          5.0         3.6 setosa
## 6          5.4         3.9 setosa
```

1.1.11 Filter Rows: Columns Meet Criteria (filter_at)

Use filter_at to find rows which meet some criteria such as maximum:

```
new.mtcars <- mtcars %>% filter_at(vars(cyl, hp),
  all_vars(. == max(.)))
new.mtcars
```

```
##                 mpg cyl disp  hp drat   wt qsec vs am gear carb
## Maserati Bora   15   8  301 335 3.54 3.57 14.6  0  1    5    8
```

Note that only one car, the Maserati Bora, had both the maximum number of cylinders and the maximum horsepower for each column, respectively.

Another example dataset comes from Suzan Baert's blog (https://suzan.rbind. io/2018/02/dplyr-tutorial-3/#filter-at), using sleep study research.

Load the msleep dataframe from the package ggplot2:

```
msleep <- ggplot2::msleep
msleep
```

```
## # A tibble: 83 x 11
##      name   genus vore  order conservation sleep_total sleep_rem sleep_
cycle awake
##      <chr> <chr> <chr> <chr> <chr>                <dbl>      <dbl>     <dbl> <dbl>
##  1 Chee~ Acin~ carni Carn~ lc                    12.1       NA        NA   11.9
##  2 Owl ~ Aotus omni  Prim~ <NA>                  17         1.8       NA   7
##  3 Moun~ Aplo~ herbi Rode~ nt                    14.4       2.4       NA   9.6
##  4 Grea~ Blar~ omni  Sori~ lc                    14.9       2.3     0.133  9.1
##  5 Cow   Bos   herbi Arti~ domesticated           4         0.7     0.667 20
##  6 Thre~ Brad~ herbi Pilo~ <NA>                  14.4       2.2     0.767  9.6
##  7 Nort~ Call~ carni Carn~ vu                     8.7       1.4     0.383 15.3
##  8 Vesp~ Calo~ <NA>  Rode~ <NA>                   7         NA        NA   17
##  9 Dog   Canis carni Carn~ domesticated          10.1       2.9     0.333 13.9
## 10 Roe ~ Capr~ herbi Arti~ lc                     3         NA        NA   21
## # ... with 73 more rows, and 2 more variables: brainwt <dbl>, bodywt <dbl>
msleep.over.5 <- msleep %>%
  select(name, sleep_total:sleep_rem, brainwt:bodywt) %>%
  filter_at(vars(contains("sleep")), all_vars(.>5))
msleep.over.5
```

```
## # A tibble: 2 x 5
##   name                sleep_total sleep_rem brainwt bodywt
##   <chr>                     <dbl>     <dbl>   <dbl>  <dbl>
## 1 Thick-tailed opposum       19.4       6.6      NA   0.37
## 2 Giant armadillo            18.1       6.1   0.081  60
```

For the preceding code, ignore the select statement for the moment (covered later). The filter_at function says to look at only variables containing the word "sleep." Within those variables (in this case, two of them), filter for *any* values greater than 5. The "" means any variable with sleep in the name. Only two rows met the criteria for the filter in this case.

1.2 Arrange (Sort)

Arrange, the sorting function, is as old as the alphabet. Based on the defined ASCII order, it rearranges a dataframe or vector in a sequence defined as either ascending or descending. Sort keys are defined as primary, secondary, and so on.

Load the msleep dataframe from the package ggplot2:

```
msleep <- ggplot2::msleep
msleep[,1:4]

## # A tibble: 83 x 4
##     name                       genus      vore  order
##     <chr>                      <chr>      <chr> <chr>
##  1 Cheetah                     Acinonyx   carni Carnivora
##  2 Owl monkey                  Aotus      omni  Primates
##  3 Mountain beaver             Aplodontia herbi Rodentia
##  4 Greater short-tailed shrew  Blarina    omni  Soricomorpha
##  5 Cow                         Bos        herbi Artiodactyla
##  6 Three-toed sloth            Bradypus   herbi Pilosa
##  7 Northern fur seal           Callorhinus carni Carnivora
##  8 Vesper mouse                Calomys    <NA>  Rodentia
##  9 Dog                         Canis      carni Carnivora
## 10 Roe deer                    Capreolus  herbi Artiodactyla
## # ... with 73 more rows
```

1.2.1 Ascending

```
animal.name.sequence <- arrange(msleep, vore, order)
animal.name.sequence[,1:4]
```

```
## # A tibble: 83 x 4
##    name                genus       vore  order
##    <chr>               <chr>       <chr> <chr>
##  1 Cheetah             Acinonyx    carni Carnivora
##  2 Northern fur seal   Callorhinus carni Carnivora
##  3 Dog                 Canis       carni Carnivora
##  4 Domestic cat        Felis       carni Carnivora
##  5 Gray seal           Halichoerus carni Carnivora
##  6 Tiger               Panthera    carni Carnivora
##  7 Jaguar              Panthera    carni Carnivora
##  8 Lion                Panthera    carni Carnivora
##  9 Caspian seal        Phoca       carni Carnivora
## 10 Genet               Genetta     carni Carnivora
## # ... with 73 more rows
```

1.2.2 Descending

```
animal.name.sequence.desc <- arrange(msleep, vore, desc(order))
head(animal.name.sequence.desc[,1:4])
```

```
## # A tibble: 6 x 4
##    name                      genus         vore  order
##    <chr>                     <chr>         <chr> <chr>
## 1 Northern grasshopper mouse Onychomys     carni Rodentia
## 2 Slow loris                 Nyctibeus     carni Primates
## 3 Thick-tailed opposum       Lutreolina    carni Didelphimorphia
## 4 Long-nosed armadillo       Dasypus       carni Cingulata
## 5 Pilot whale                Globicephalus carni Cetacea
## 6 Common porpoise            Phocoena      carni Cetacea
```

In section "Mutate," you'll see how a variable can be created on the fly and then used in the same statement for sorting.

1.3 Rename

Rename allows you to change the name of one or more columns. It is a convenience function and changes no data.

Rename one or more columns in a dataset:

```
names(iris)
```

```
## [1] "Sepal.Length" "Sepal.Width"  "Petal.Length" "Petal.
Width"  "Species"
```

Show new column names:

```
renamed.iris <- rename(iris, width.of.petals = Petal.Width,
various.plants.and.animals = Species)
names(renamed.iris)
```

```
## [1] "Sepal.Length"                "Sepal.Width"
## [3] "Petal.Length"                "width.of.petals"
## [5] "various.plants.and.animals"
```

1.4 Mutate

Mutate adds new variables to a dataframe. It requires the original dataframe as the first argument and then arguments to create new variables as the remaining arguments. The following example adds the natural log of length and weight to the dataframe created earlier that contains just the length and weight variables.

Add a new, calculated variable to a dataframe:

```
data("ChickWeight")
ChickWeight[1:2,]  #first two rows
```

```
##   weight Time Chick Diet
## 1     42    0     1    1
## 2     51    2     1    1
```

First two rows, with new field added:

```
Chickweight.with.log <- mutate(ChickWeight,
log.of.weight = log10(weight))
Chickweight.with.log[1:2,]
```

```
##   weight Time Chick Diet log.of.weight
## 1     42    0     1    1      1.623249
## 2     51    2     1    1      1.707570
```

1.4.1 mutate_all to Add New Fields All at Once

Create a series of new fields based on some calculation using existing fields. This is a shorthand approach, making it unnecessary to specifically define each new variable. In the following example, each of the designated columns (6–11) supplies numbers to the square root function and then creates a new column with the suffix "_square root."

Load the msleep dataframe from the package ggplot2:

```
msleep <- ggplot2::msleep
names(msleep)
```

```
##  [1] "name"         "genus"       "vore"         "order"   "conservation"
##  [6] "sleep_total"  "sleep_rem"   "sleep_cycle"  "awake"   "brainwt"
## [11] "bodywt"
```

Now add square root variables/columns for the numeric variables in columns 6–11:

```
msleep.with.square.roots <- mutate_all(msleep[,6:11],
  funs("square root" = sqrt( . )))
names(msleep.with.square.roots)
```

You may get a warning message on R 4.0.3 using DPLYR > 0.8.0.

```
##  [1] "sleep_total"             "sleep_rem"
##  [3] "sleep_cycle"             "awake"
##  [5] "brainwt"                 "bodywt"
##  [7] "sleep_total_square root" "sleep_rem_square root"
##  [9] "sleep_cycle_square root" "awake_square root"
## [11] "brainwt_square root"     "bodywt_square root"
```

```
msleep.with.square.roots
```

```
## # A tibble: 83 x 12
##   sleep_total sleep_rem sleep_cycle awake brainwt bodywt `sleep_total_sq~
##         <dbl>     <dbl>       <dbl> <dbl>   <dbl>  <dbl>            <dbl>
##  1        12.1        NA          NA  11.9 NA          50             3.48
##  2        17         1.8          NA   7    0.0155     0.48           4.12
##  3        14.4       2.4          NA   9.6 NA           1.35          3.79
##  4        14.9       2.3       0.133   9.1  0.00029     0.019         3.86
##  5         4         0.7       0.667  20    0.423     600             2
##  6        14.4       2.2       0.767   9.6 NA           3.85          3.79
##  7         8.7       1.4       0.383  15.3 NA          20.5           2.95
##  8         7          NA          NA  17   NA           0.045         2.65
##  9        10.1       2.9       0.333  13.9  0.07       14             3.18
## 10         3          NA          NA  21    0.0982     14.8           1.73
## # ... with 73 more rows, and 5 more variables: `sleep_rem_square root` <dbl>,
## #   `sleep_cycle_square root` <dbl>, `awake_square root` <dbl>,
`brainwt_square
## #   root` <dbl>, `bodywt_square root` <dbl>
```

1.4.2 mutate_at to Add Fields

Variables for Class-rank, Age-rank, and Survived-rank are created as follows, using the Titanic survivor dataset. Using the data function, Titanic loads as a table; as.data.frame is used to convert to a dataframe:

```
data("Titanic")
Titanic <- as.data.frame(Titanic)
head(Titanic)
```

```
##   Class    Sex   Age Survived Freq
## 1   1st   Male Child       No    0
## 2   2nd   Male Child       No    0
## 3   3rd   Male Child       No   35
## 4  Crew   Male Child       No    0
## 5   1st Female Child       No    0
## 6   2nd Female Child       No    0
```

```
titanic.with.ranks <- mutate_at(Titanic, vars(Class,Age,Survived),
funs(Rank = min_rank(desc(.))))
head(titanic.with.ranks)
```

##	Class	Sex	Age	Survived	Freq	Class_Rank	Age_Rank	Survived_Rank
## 1	1st	Male	Child	No	0	25	17	17
## 2	2nd	Male	Child	No	0	17	17	17
## 3	3rd	Male	Child	No	35	9	17	17
## 4	Crew	Male	Child	No	0	1	17	17
## 5	1st	Female	Child	No	0	25	17	17
## 6	2nd	Female	Child	No	0	17	17	17

1.4.3 mutate_if

mutate_if combines IF logic and mutate to create a new variable or alter an existing one.[1] The following are two examples. In example one, factors are removed for later convenience. Example two shows how NAs in a column can be replaced by zero.

Example #1

Create a simple function to divide a number by 10:

```
divide.by.10 <- function (a.number) (a.number / 10)
```

Use the built-in dataset CO2 to illustrate:

```
head(CO2)
```

##	Plant	Type	Treatment	conc	uptake
## 1	Qn1	Quebec	nonchilled	95	16.0
## 2	Qn1	Quebec	nonchilled	175	30.4
## 3	Qn1	Quebec	nonchilled	250	34.8
## 4	Qn1	Quebec	nonchilled	350	37.2
## 5	Qn1	Quebec	nonchilled	500	35.3
## 6	Qn1	Quebec	nonchilled	675	39.2

---

[1] "TidyVerse – mutate_at(), mutate_if(), mutate_all()," accessed on January 20, 2021, http:// rstudio-pubs-static.s3.amazonaws.com/491966_5072cd8affd84069833a65aed85303b8.html.

Now divide any column which has a numeric format by 10, using the preceding custom function:

```
new.df <- CO2 %>%
  mutate_if(is.numeric, divide.by.10)
head(new.df)
```

```
##   Plant   Type  Treatment conc uptake
## 1   Qn1 Quebec nonchilled  9.5   1.60
## 2   Qn1 Quebec nonchilled 17.5   3.04
## 3   Qn1 Quebec nonchilled 25.0   3.48
## 4   Qn1 Quebec nonchilled 35.0   3.72
## 5   Qn1 Quebec nonchilled 50.0   3.53
## 6   Qn1 Quebec nonchilled 67.5   3.92
```

Any NA in a numeric field is replaced by zero. "coalesce" is a function of DPLYR which finds the first non-missing value at each position. A set of three dots in R (and other languages) is called an ellipsis. It means that a function will accept any number of arguments:

```
df <- data.frame(
alpha = c(22, 1, NA),
almond = c(0, 5, 10),
grape = c(0, 2, 2),
apple = c(NA, 5, 10))
df
```

```
##    alpha almond grape apple
## 1    22      0     0    NA
## 2     1      5     2     5
## 3    NA     10     2    10
df.fix.alpha <- df %>% mutate_if(is.numeric, coalesce, ... = 0)
df.fix.alpha
```

```
##    alpha almond grape apple
## 1    22      0     0     0
## 2     1      5     2     5
## 3     0     10     2    10
```

1.4.4 String Detect and True/False Duplicate Indicator

All columns with the characters "sleep" in the column name are converted to character.

Load the msleep dataframe from the package ggplot2:

```
msleep <- ggplot2::msleep
table(msleep$vore)

##    carni   herbi insecti    omni
##       19      32       5      20

msleep.no.c.or.a <- filter(msleep, !str_detect(vore,
    paste(c("c","a"), collapse = "|")))
table(msleep.no.c.or.a$vore)

##
## herbi   omni
##    32     20
```

Add a field indicating whether a particular value in a column occurs more than once. In this case, the column is conservation:

```
msleep.with.dup.indicator <- mutate(msleep, duplicate.indicator =
duplicated(conservation))
msleep.with.dup.indicator[1:6,]

## # A tibble: 6 x 12
##     name  genus vore  order conservation sleep_total sleep_rem sleep_cycle
awake
##    <chr> <chr> <chr> <chr> <chr>               <dbl>     <dbl>       <dbl>    <dbl>
## 1 Chee~ Acin~ carni Carn~ lc                   12.1        NA          NA    11.9
## 2 Owl ~ Aotus omni  Prim~ <NA>                 17          1.8         NA     7
## 3 Moun~ Aplo~ herbi Rode~ nt                   14.4        2.4         NA     9.6
## 4 Grea~ Blar~ omni  Sori~ lc                   14.9        2.3       0.133    9.1
## 5 Cow   Bos   herbi Arti~ domesticated          4          0.7       0.667   20
## 6 Thre~ Brad~ herbi Pilo~ <NA>                 14.4        2.2       0.767    9.6
## # ... with 3 more variables: brainwt <dbl>, bodywt <dbl>,
## #   duplicate.indicator <lgl>
```

Note the new field, duplicate.indicator, shown at the bottom. Mutate adds a field indicating whether a particular value in a column occurs more than once. In this case, the column is conservation. Figure 1-1 shows a partial view of msleep.with.dup.indicator, with the duplicate indicator field clearly shown.

sleep_total	sleep_rem	sleep_cycle	awake	brainwt	bodywt	duplicate.indica
12.1	NA	NA	11.90	NA	50.000	FALSE
17.0	1.8	NA	7.00	0.01550	0.480	FALSE
14.4	2.4	NA	9.60	NA	1.350	FALSE
14.9	2.3	0.1333333	9.10	0.00029	0.019	TRUE
4.0	0.7	0.6666667	20.00	0.42300	600.000	FALSE
14.4	2.2	0.7666667	9.60	NA	3.850	TRUE
8.7	1.4	0.3833333	15.30	NA	20.490	FALSE
7.0	NA	NA	17.00	NA	0.045	TRUE
10.1	2.9	0.3333333	13.90	0.07000	14.000	TRUE
3.0	NA	NA	21.00	0.09820	14.800	TRUE
5.3	0.6	NA	18.70	0.11500	33.500	TRUE
9.4	0.8	0.2166667	14.60	0.00550	0.728	TRUE
10.0	0.7	NA	14.00	NA	4.750	TRUE

Figure 1-1. *Created field, duplicate.indicator, shown on the right*

```
msleep.with.dup.indicator <- mutate(msleep,
  duplicate.indicator = duplicated(conservation))
msleep.with.dup.indicator[1:6,c(1,2,3,12)]

## # A tibble: 6 x 4
##   name                         genus      vore  duplicate.indicator
##   <chr>                        <chr>      <chr> <lgl>
## 1 Cheetah                      Acinonyx   carni FALSE
## 2 Owl monkey                   Aotus      omni  FALSE
## 3 Mountain beaver              Aplodontia herbi FALSE
## 4 Greater short-tailed shrew   Blarina    omni  TRUE
## 5 Cow                          Bos        herbi FALSE
## 6 Three-toed sloth             Bradypus   herbi TRUE
```

Sort by conservation as major key and genus as minor key:

```
msleep.with.dup.indicator2 <- mutate(msleep,
  duplicate.indicator =  duplicated(conservation, genus)) %>%
    arrange(conservation,genus)
msleep.with.dup.indicator2
```

```
## # A tibble: 83 x 12
##      name  genus vore  order conservation sleep_total sleep_rem sleep_
cycle awake
##      <chr> <chr> <chr> <chr> <chr>              <dbl>     <dbl>   <dbl> <dbl>
##  1 Gira~ Gira~ herbi Arti~ cd                   1.9       0.4      NA   22.1
##  2 Pilo~ Glob~ carni Ceta~ cd                   2.7       0.1      NA   21.4
##  3 Cow   Bos   herbi Arti~ domesticated         4         0.7   0.667   20
##  4 Dog   Canis carni Carn~ domesticated        10.1       2.9   0.333   13.9
##  5 Guin~ Cavis herbi Rode~ domesticated         9.4       0.8   0.217   14.6
##  6 Chin~ Chin~ herbi Rode~ domesticated        12.5       1.5   0.117   11.5
##  7 Horse Equus herbi Peri~ domesticated         2.9       0.6   1       21.1
##  8 Donk~ Equus herbi Peri~ domesticated         3.1       0.4      NA   20.9
##  9 Dome~ Felis carni Carn~ domesticated        12.5       3.2   0.417   11.5
## 10 Rabb~ Oryc~ herbi Lago~ domesticated         8.4       0.9   0.417   15.6
## # ... with 73 more rows, and 3 more variables: brainwt <dbl>, bodywt
<dbl>,
## #   duplicate.indicator <lgl>
```

Both conservation and genus have to be duplicated for the duplicate.indicator to be set to TRUE.

Create toy[2] dataframe:

```
fruit <- c("apple","pear","orange","grape", "orange","orange")
x <- c(1,2,4,9,4,6)
y <- c(22,3,4,55,15,9)
z <- c(3,1,4,10,12,8)
w <- c(2,2,2,4,5,6)
```

---

[2]The word "toy" in this context simply means a tiny dataset used solely to demonstrate a section of code.

```
df <- data.frame(fruit,x,y,z,w)
df
```

```
##     fruit x  y   z w
## 1   apple 1 22   3 2
## 2    pear 2  3   1 2
## 3  orange 4  4   4 2
## 4   grape 9 55  10 4
## 5  orange 4 15  12 5
## 6  orange 6  9   8 6
```

```
df.show.single.dup <- mutate(df, duplicate.indicator = duplicated(fruit))
df.show.single.dup
```

```
##     fruit x  y   z w duplicate.indicator
## 1   apple 1 22   3 2                FALSE
## 2    pear 2  3   1 2                FALSE
## 3  orange 4  4   4 2                FALSE
## 4   grape 9 55  10 4                FALSE
## 5  orange 4 15  12 5                 TRUE
## 6  orange 6  9   8 6                 TRUE
```

DPLYR goes down the rows, looking at the values in the column "fruit." It cannot detect the first duplicate because it does not yet know if there is another one. When it sees "orange" a second and third time, it sets the duplicate.indicator column as "TRUE" for the rows containing the value "orange."

If you want to check for multiple fields, use mutate to combine them into a new, single field before using the duplicate logical (True/False) indicator.

1.4.5 Drop Variables Using NULL

```
fruit <- c("apple","pear","orange","grape", "orange","orange")
x <- c(1,2,4,9,4,6)
y <- c(22,3,4,55,15,9)
z <- c(3,1,4,10,12,8)
df <- data.frame(fruit,x,y,z)
df <- mutate(df, z = NULL)
df
```

```
##      fruit x  y
## 1    apple 1 22
## 2     pear 2  3
## 3   orange 4  4
## 4    grape 9 55
## 5   orange 4 15
## 6   orange 6  9
```

1.4.6 Preferred coding sequence

John Mount, in his September 22, 2017, blog (www.r-bloggers.com/my-advice-on-dplyrmutate/), suggests using multiple mutate statements to create variables rather than putting them all in one place. At one point, this made a difference when using certain database files but has subsequently been corrected. Nonetheless, the multiple mutate format seems safer and more intuitive. The following shows not recommended and recommended coding methods.

Not recommended but works:

```
if (!require("nycflights13")) install.packages("nycflights13")

mutate(flights,
  gain = arr_delay - dep_delay,
  hours = air_time / 60,
  gain_per_hour = gain / hours,
  gain_per_minute = 60 * gain_per_hour)
```

```
## A tibble: 336,776 x 23
##   year month   day dep_time sched_dep_time dep_delay arr_time sched_arr_time
##   <int> <int> <int>  <int>      <int>         <dbl>    <int>       <int>
## 1  2013     1     1    517        515           2      830         819
## 2  2013     1     1    533        529           4      850         830
## 3  2013     1     1    542        540           2      923         850
## 4  2013     1     1    544        545          -1     1004        1022
## 5  2013     1     1    554        600          -6      812         837
## 6  2013     1     1    554        558          -4      740         728
## 7  2013     1     1    555        600          -5      913         854
## 8  2013     1     1    557        600          -3      709         723
```

```
##  9 2013    1    1    557    600    -3    838    846
## 10 2013    1    1    558    600    -2    753    745
##  ... with 336,766 more rows, and 15 more variables: arr_delay <dbl>,
##  carrier <chr>, flight <int>, tailnum <chr>, origin <chr>, dest <chr>,
##  air_time <dbl>, distance <dbl>, hour <dbl>, minute <dbl>, time_hour
<dttm>,
##  gain <dbl>, hours <dbl>, gain_per_hour <dbl>, gain_per_minute <dbl>
```

Recommended:

```
if (!require("nycflights13")) install.packages("nycflights13")
newfield.flights <- flights %>%
  mutate(gain = arr_delay - dep_delay,
  hours = air_time / 60) %>%
  mutate(gain_per_hour = gain / hours) %>%
  mutate(gain_per_minute = 60 * gain_per_hour)
```

Show selected columns for the first six rows, including the newly created ones.

```
newfield.flights[1:6,c(1:2,20:23)]
```

```
## A tibble: 6 x 6
##
```

year month gain hours gain_per_hour gain_per_minute

	year	month	gain	hours	gain_per_hour	gain_per_minute
##	<int>	<int>	<dbl>	<dbl>	<dbl>	<dbl>
## 1	2013	1	9	3.78	2.38	143.
## 2	2013	1	16	3.78	4.23	254.
## 3	2013	1	31	2.67	11.6	698.
## 4	2013	1	-17	3.05	-5.57	-334.
## 5	2013	1	-19	1.93	-9.83	-590.
## 6	2013	1	16	2.5	6.4	384

```
# A tibble: 6 x 6
# year month gain hours gain_per_hour gain_per_minute
```

1.4.7 Transmute: Keep Only Variables Created

Transmute allows you to create an entirely new dataframe based on calculations performed on existing variables:

```
fruit <- c("apple","pear","orange","grape", "orange","orange")
x <- c(1,2,4,9,4,6)
y <- c(22,3,4,55,15,9)
z <- c(3,1,4,10,12,8)
df <- data.frame(fruit,x,y,z)
df #before transmute

##     fruit x  y  z
## 1   apple 1 22  3
## 2    pear 2  3  1
## 3 orange 4  4  4
## 4   grape 9 55 10
## 5 orange 4 15 12
## 6 orange 6  9  8
df <- transmute(df, new.variable = x + y + z)

#row "apple" =1 + 22 + 3 = 26
#row "pear" = 2 + 3 + 1 = 6 and so on
```

After transmute:

```
df

##   new.variable
## 1           26
## 2            6
## 3           12
## 4           74
## 5           31
## 6           23
```

1.4.8 Use Across to Apply a Function over Multiple Columns

Define a simple function to double the value of a number:

```
double.it <- function(x) x*2
```

Original iris:

```
head(iris)
```

```
##   Sepal.Length Sepal.Width Petal.Length Petal.Width Species
## 1          5.1         3.5          1.4         0.2  setosa
## 2          4.9         3.0          1.4         0.2  setosa
## 3          4.7         3.2          1.3         0.2  setosa
## 4          4.6         3.1          1.5         0.2  setosa
## 5          5.0         3.6          1.4         0.2  setosa
## 6          5.4         3.9          1.7         0.4  setosa
```

Show a new iris dataframe with doubled values for the numeric columns:

```
iris %>%
  mutate(across(where(is.numeric), double.it)) %>%
  head()
```

```
##   Sepal.Length Sepal.Width Petal.Length Petal.Width Species
## 1         10.2         7.0          2.8         0.4  setosa
## 2          9.8         6.0          2.8         0.4  setosa
## 3          9.4         6.4          2.6         0.4  setosa
## 4          9.2         6.2          3.0         0.4  setosa
## 5         10.0         7.2          2.8         0.4  setosa
## 6         10.8         7.8          3.4         0.8  setosa
```

1.4.9 Conditional Mutating Using case_when

Using conditional mutating plus case_when, you can mutate a new field and then set values based on multiple conditions.[3]

```
row1 <- c("a","b","c","d","e","f","column.to.be.changed")
row2 <- c(1,1,1,6,6,1,2)
row3 <- c(3,4,4,6,4,4,4)
row4 <- c(4,6,25,5,5,2,9)
row5 <- c(5,3,6,3,3,6,2)
df <- as.data.frame(rbind(row2,row3,row4,row5))
names(df) <- row1

df

##       a b  c d e f column.to.be.changed
## row2 1 1  1 6 6 1                    2
## row3 3 4  4 6 4 4                    4
## row4 4 6 25 5 5 2                    9
## row5 5 3  6 3 3 6                    2
new.df <-df %>%
  mutate(column.to.be.changed = case_when(a == 2 | a == 5 |
  a == 7 | (a == 1 & b == 4) ~ 2, a == 0 | a == 1 | a == 4 |
  a == 3 | c == 4 ~ 3, TRUE ~ NA_real_))
```

This is a series of "OR" conditions. If any of them are true, then the last column (column.to.be.changed) will be either a 2 or a 3.

Modified:

```
new.df

##   a b  c d e f column.to.be.changed
## 1 1 1  1 6 6 1                    3
## 2 3 4  4 6 4 4                    3
## 3 4 6 25 5 5 2                    3
## 4 5 3  6 3 3 6                    2
```

---

[3]https://stackoverflow.com/questions/24459752/can-dplyr-package-be-used-for-conditional-mutating. Accessed on January 20, 2021.

1.5 Select to Choose Variables/Columns

Select retains only the variables you include in the select statement. The rename function, usually mentioned as a sister function of select, does not drop any variables.

Be aware that other packages, such as Zeilig, use the select command. You'll pull your hair out trying to figure out R's error messages if you write syntactically correct code based on DPLYR/tidyverse but have loaded some other package, which uses the same command, last. There are two ways to get around this:

- Always enter "library(tidyverse)" *last*, after all other packages have been loaded.

- Explicitly tie the command to the package: "dplyr::select". Keep in mind that DPLYR is part of the tidyverse collection of coordinated packages.

1.5.1 Delete a Column

```
fruit <- c("apple","pear","orange","grape", "orange","orange")
x <- c(1,2,4,9,4,6)
y <- c(22,3,4,55,15,9)
z <- c(3,1,4,10,12,8)
df <- data.frame(fruit,x,y,z) #before select
df

##      fruit x  y  z
## 1   apple 1 22  3
## 2    pear 2  3  1
## 3 orange 4  4  4
## 4   grape 9 55 10
## 5 orange 4 15 12
## 6 orange 6  9  8
```

Put a minus sign in front of any variable(s) to be dropped. In this case, the fruit column is not included in the new dataframe:

```
new.df.no.fruit <- dplyr::select(df, -fruit)
new.df.no.fruit #after select
```

```
##   x  y  z
## 1 1 22  3
## 2 2  3  1
## 3 4  4  4
## 4 9 55 10
## 5 4 15 12
## 6 6  9  8
```

1.5.2 Delete Columns by Name Using starts_with or ends_with

Use the names command to list column names in the dataframe:

```
data("mtcars")
names(mtcars)
```

```
##  [1] "mpg"  "cyl"  "disp" "hp"   "drat" "wt"  "qsec" "vs"  "am"  "gear"
## [11] "carb"
```

Delete any columns with names starting with a "d":

```
mtcars.no.col.names.start.with.d <- select(mtcars, -starts_with("d"))
names(mtcars.no.col.names.start.with.d)
```

```
## [1] "mpg"  "cyl"  "hp"   "wt"   "qsec" "vs"  "am"  "gear" "carb"
```

Drop any columns with names ending in "t". In this case, "drat" and "wt" are removed:

```
mtcars.no.col.names.ends.with <- select(mtcars,
   - ends_with("t"))
names(mtcars.no.col.names.ends.with)
```

```
## [1] "mpg"  "cyl"  "disp" "hp"   "qsec" "vs"  "am"   "gear" "carb"
```

1.5.3 Rearrange Column Order

```
fruit <- c("apple","pear","orange","grape", "orange","orange")
x <- c(1,2,4,9,4,6)
y <- c(22,3,4,55,15,9)
z <- c(3,1,4,10,12,8)
df <- data.frame(fruit,x,y,z)
df

##     fruit x  y  z
## 1  apple 1 22  3
## 2   pear 2  3  1
## 3 orange 4  4  4
## 4  grape 9 55 10
## 5 orange 4 15 12
## 6 orange 6  9  8
```

   This action moves column z to the left of everything else and thus becomes the first column. It is often more convenient to have frequently used columns on the left. The keyword "everything" means that all remaining columns should be retained.

1.5.4 select_all to Apply a Function to All Columns

Use select_all to apply a function to all columns.
   Before - column names are not capitalized:

```
#create new dataframe
state <- c("Maryland", "Alaska", "New Jersey")
income <- c(76067,74444,73702)
median.us <- c(61372,61372,61372)
life.expectancy <- c(78.8,78.3,80.3)
top.3.states <- data.frame(state, income, median.us,
  life.expectancy)

top.3.states #before - column names are not capitalized
```

```
##          state income median.us life.expectancy m.some.number
## 1    Maryland  76067     61372            78.8            33
## 2      Alaska  74444     61372            78.3            11
## 3 New Jersey  73702     62372            80.3            44
```

Capitalize column names, using the "toupper" function:

```
new.top.3.states <- select_all(top.3.states, toupper)
new.top.3.states #after function "toupper" applied
```

```
##          STATE INCOME MEDIAN.US LIFE.EXPECTANCY M.SOME.NUMBER
## 1    Maryland  76067     61372            78.8            33
## 2      Alaska  74444     61372            78.3            11
## 3 New Jersey  73702     62372            80.3            44
```

1.5.5 Select Columns Using the Pull Function

The pull function acts somewhat like the dataframe$variable syntax in R. It isolates a specified column. You can specify a column from the left (e.g., second column via 2) or, using a negative number, a column from the right.

```
top.3.states <- data.frame(state, income, median.us, life.expectancy)
top.3.states #display dataframe values
```

```
##          state income median.us life.expectancy
## 1    Maryland  76067     61372            78.8
## 2      Alaska  74444     61372            78.3
## 3 New Jersey  73702     61372            80.3
```

Get only the first column from the left, the state:

```
pull.first.column <- pull(top.3.states,1)
pull.first.column
```

```
## [1] "Maryland"   "Alaska"     "New Jersey"
```

Use a negative number to pull a column from the right. -1 = rightmost column:

```
pull.last.column <- pull(top.3.states,-1)
pull.last.column
```

```
## [1] 78.8 78.3 80.3
```

1.5.6 Select Rows: Any Variable Meets Some Condition

DPLYR shows its raw power in some of its less used functions. You may have a dataset where you want to know, for example, if anything exceeds a certain value. The following code filters for anything in the dataset over 200, in any column.

Number of rows in original dataframe:

```
nrow(mtcars)
```

```
## [1] 32
```

Show anything, anywhere, more than 200:

```
mtcars.more.than.200 <- filter_all(mtcars, any_vars(. > 200))
nrow(mtcars.more.than.200)
```

```
## [1] 16
```

1.5.7 Select Columns: Omit If Column Name Contains Specific Characters

Select specified columns plus any column without a "p":

```
names(mtcars)
```

```
##  [1] "mpg"  "cyl"  "disp" "hp"   "drat" "wt"  "qsec" "vs"  "am"   "gear"
## [11] "carb"
```

Names of dataframe columns after select statement:

```
cars.with.no.p <- mtcars %>%
  dplyr::select(-contains("p"))
names(cars.with.no.p)
```

```
## [1] "cyl"  "drat" "wt"   "qsec" "vs"   "am"   "gear" "carb"
```

No column name with a "p" anywhere is included in the new dataframe.

1.5.8 Select Using Wildcard Matching

Show column names of the built-in mtcars dataframe:

```
names(mtcars)
```

```
##  [1] "mpg"  "cyl" "disp" "hp"   "drat" "wt" "qsec" "vs" "am" "gear"
## [11] "carb"
```

Column names selected contain the characters "pg" or "gea". The pipe symbol means "OR":

```
subset.mtcars <- select(mtcars,
  matches("pg|gea"))
names(subset.mtcars)
```

```
## [1] "mpg"   "gear"
```

The function "matches" is more general than "contains" because it is a regular expression and therefore more flexible.

1.6 Joins: Manipulations of Data from Two Sources

According to Wickham and Grolemund,[4] when working with relational data, there are a set of verbs needed to work with pairs of tables. These verbs fall into three families:

- Mutating joins, which add new variables to one data from matching observations in another

- Filtering joins, which filter observations from one dataframe based on whether or not they match an observation in the other table

- Set operations, which treat observations as if they were set elements

---

[4]"13 Relational Data | R for Data Science," accessed on January 20, 2021, `https://r4ds.had.co.nz/relational-data.html`.

Connections between tables are made using defined keys. Consider two simple tables: Table A contains a US state abbreviation and that state's population. Table B contains the US state abbreviation and the full US state name. You want to create a new dataframe which has US state full name and state population. In table A, state abbreviation is a primary key. The corresponding state abbreviation in table B is a foreign key.

The following are examples of more frequently used joins.

1.6.1 Left Join (Most Common)

Left join: match both files using key (by = "key"); keep all records on the left dataframe for any matching records on the right dataframe, and add data to a new column to create the output dataframe:

```
us.state.areas <- as.data.frame(cbind(state.abb, state.area))
us.state.areas[1:3,]
```

```
##    state.abb state.area
## 1         AL      51609
## 2         AK     589757
## 3         AZ     113909
us.state.abbreviation.and.name <- as.data.frame(cbind(state.abb,
  state.name))
us.state.abbreviation.and.name[1:3,]
```

```
##    state.abb state.name
## 1         AL    Alabama
## 2         AK     Alaska
## 3         AZ    Arizona
```

```
state.info.abb.area.name <- us.state.areas %>%
  left_join(us.state.abbreviation.and.name, by = "state.abb")
head(state.info.abb.area.name)
```

```
##    state.abb state.area state.name
## 1         AL      51609    Alabama
## 2         AK     589757     Alaska
## 3         AZ     113909    Arizona
```

```
## 4        AR      53104   Arkansas
## 5        CA      158693 California
## 6        CO      104247  Colorado
```

1.6.2 Inner Join

The inner join function outputs only the rows in both dataframes when the keys are the same. All other rows are dropped. In the following example, team.info and school.and. team have Sally, Tom, and Alfonzo as common keys. Frieda and Bill are dropped from the output.

Create first dataframe:

```
names <- c("Sally","Tom","Frieda","Alfonzo")
team.scores <- c(3,5,2,7)
team.league <- c("alpha","beta","gamma", "omicron")
team.info <- data.frame(names, team.scores, team.league)
```

Create second dataframe:

```
names = c("Sally","Tom", "Bill", "Alfonzo")
school.grades <- c("A","B","C","B")
school.info <- data.frame(names, school.grades)
school.and.team <- inner_join(team.info, school.info, by = "names")
school.and.team
```

```
##      names team.scores team.league school.grades
## 1    Sally           3       alpha             A
## 2      Tom           5        beta             B
## 3  Alfonzo           7     omicron             B
```

Data appears on the school.and.team dataframe only when names match exactly.

1.6.3 Anti-join

Anti-join outputs x when there is no match with y. In the following example, anti-join keeps all values from team.info with *no match* in school.info. Pay attention to the order in which you list the two dataframes because switching them will change the outcome. "Names" is the key in this case.

An example use of anti-join is matching a human resources table with a payroll table in an ERP system. For example, if you match by employee number and find that an employee is on the human resources file but not on the payroll file, then you can assume the employee has not yet been set up in payroll.

Create first dataframe:

```
names <- c("Sally","Tom","Frieda","Alfonzo")
team.scores <- c(3,5,2,7)
team.league <- c("alpha","beta","gamma", "omicron")
team.info <- data.frame(names, team.scores, team.league)
team.info
##       names team.scores team.league
## 1    Sally           3        alpha
## 2      Tom           5         beta
## 3   Frieda           2        gamma
## 4  Alfonzo           7      omicron
```

Create second dataframe:

```
names <- c("Sally","Tom", "Bill", "Alfonzo")
school.grades <- c("A","B","C","B")
school.info <- data.frame(names, school.grades)
school.info

##       names school.grades
## 1    Sally             A
## 2      Tom             B
## 3     Bill             C
## 4  Alfonzo             B
```

Data is from team.info, but only names NOT matching grade data are shown. Frieda has no grades:

```
team.info.but.no.grades <- anti_join(team.info, school.info,
   by = "names")

team.info.but.no.grades

##     names team.scores team.league
## 1 Frieda           2        gamma
```

1.6.4 Full Join

This matching function keeps all values from *both* dataframes. Note that columns where data does not exist show row values of NA. For example, Bill is contained in school.info but not in team.info. So his name and grade show up, but he has no team score or team league.

Create first dataframe:

```
names = c("Sally","Tom","Frieda","Alfonzo")
team.scores = c(3,5,2,7)
team.league = c("alpha","beta","gamma", "omicron")
team.info = data.frame(names, team.scores, team.league)
```

Create second dataframe:

```
names = c("Sally","Tom", "Bill", "Alfonzo")
school.grades = c("A","B","C","B")
school.info = data.frame(names, school.grades)
```

Create new dataframe using full join (note that Frieda shows school.grades as "NA"):

```
team.info.and.or.grades <- full_join(team.info, school.info, by = "names")
team.info.and.or.grades
```

```
##       names team.scores team.league school.grades
## 1    Sally           3       alpha             A
## 2      Tom           5        beta             B
## 3   Frieda           2       gamma          <NA>
## 4 Alfonzo           7     omicron             B
## 5     Bill          NA        <NA>             C
```

1.6.5 Semi-join

A semi-join keeps all observations in dataset1 which match dataset2. Again, the order you list dataframes determines the outcome.

Use team.info and school.info from prior code. Keep team.info rows which have a grade:

```
team.info.with.grades <- semi_join(team.info, school.info)

## Joining, by = "names"
```

Note in the preceding code that DPLYR is smart enough to understand that "names" is a common key. Frieda has no grades so is not included in the output:

```
team.info.with.grades
```

```
##        names team.scores team.league
## 1    Sally           3         alpha
## 2      Tom           5          beta
## 3 Alfonzo           7       omicron
```

1.6.6 Right Join

The right join has the format right_join(x, y, by = common_key). It returns all rows of y and columns for both x and y dataframes. Where there is no match on x and y, the columns from x will have NAs.

If there are multiple matches between x and y, all combinations will be included in the output.

```
us.state.areas <- as.data.frame(cbind(state.abb, state.area))
us.state.areas[1:3,]
```

```
##    state.abb state.area
## 1         AL      51609
## 2         AK     589757
## 3         AZ     113909
us.state.abbreviation.and.name <- as.data.frame(cbind(state.abb,
   state.name))
us.state.abbreviation.and.name[1:3,]
```

```
##    state.abb state.name
## 1         AL    Alabama
## 2         AK     Alaska
## 3         AZ    Arizona
```

In this example, Alabama is replaced by "intentional mismatch." As a result, it is missing in the output. However, all columns from both datasets are included:

```
us.state.abbreviation.and.name[1,1] <- "Intentional Mismatch"
us.state.with.abbreviation.and.name.and.area <- right_join(us.state.areas,
    us.state.abbreviation.and.name, by = "state.abb")
us.state.with.abbreviation.and.name.and.area[1:3,]
```

```
##    state.abb state.area state.name
## 1        AK     589757     Alaska
## 2        AZ     113909     Arizona
## 3        AR      53104     Arkansas
```

1.7 Slice

Slice provides a convenient way to include specific rows or ranges of rows in your datasets. Think of it as a row range function.

Load the msleep dataframe from the package ggplot2:

```
msleep <- ggplot2::msleep
nrow(msleep) #initially 83 rows
```

```
## [1] 83
msleep.only.first.5 <- slice(msleep, -6:-n())
```

Rows 6–83 are dropped. You do not have to know how many rows exist in the dataframe. The expression "n(")" is the total number of rows.

Now only the first five rows are retained:

```
nrow(msleep.only.first.5)
```

```
## [1] 5
msleep.20.rows <- msleep %>%
    slice(20:39)
nrow(msleep.20.rows)
```

```
## [1] 20
```

You can show the difference between the original and sliced dataframe (or tibble) as follows:

```
nrow(msleep) - nrow(msleep.20.rows)
```

```
## [1] 63
```

1.8 Summarise

Summarise is my favorite DPLYR function. Combined with group_by, it provides a treasure trove of handy, quick tools for sizing up data (counting, adding, standardm# deviation, and other functions) by one or more categorical variables.

Use the built-in dataset gehan from the MASS package. Gehan includes data related to leukemia patient remission. Unfortunately, it has a function name conflict with DPLYR, so that you need to make sure DPLYR (or better yet, tidyverse) is loaded last.

Be sure to load tidyverse AFTER loading MASS. Otherwise, you will get an error message.

Time is based on treatment or no treatment with a particular drug. For this data, longer remission times are a positive outcome for the patient.

This is a bit kludgy, but it makes the MASS-tidyverse conflict go away:

```
library(MASS)
data(gehan)
gehan2 <- gehan
library(tidyverse)
```

How many patients were in the medical trial?

```
gehan2 %>% summarise( kount = n())
```

```
##    kount
## 1     42
```

What was the count by treatment/no treatment?

```
gehan2 %>%
  group_by(treat) %>%
  summarise(kount = n())
```

```
## `summarise()` ungrouping output (override with `.groups` argument)
## # A tibble: 2 x 2
##    treat    kount
##    <fct>    <int>
## 1 6-MP         21
## 2 control      21
```

What are general statistics for treatment/no treatment?

```
gehan2 %>%
  group_by(treat) %>%
  summarise(average.remiss.time = mean(time),
    median.remiss.time = median(time),
    std.dev.remiss.time = sd(time),
    median.abs.deviation = mad(time),
     IQR.remiss.time = IQR(time))
```

```
## `summarise()` ungrouping output (override with `.groups` argument)
## # A tibble: 2 x 6
##    treat average.remiss.~ median.remiss.t~ std.dev.remiss.~ median.abs.devi~
##    <fct>          <dbl>            <int>            <dbl>            <dbl>
## 1 6-MP            17.1               16             10.0             10.4
## 2 cont~           8.67               8             6.47             5.93
## # ... with 1 more variable: IQR.remiss.time <dbl>
```

Summarise can be used to find minimum/maximum within the "by" group.

```
gehan2 %>%
  group_by(treat) %>%
  summarise(minimum.remission = min(time),
      max.remission = max(time))
```

```
## `summarise()` ungrouping output (override with `.groups` argument)
## # A tibble: 2 x 3
##   treat   minimum.remission max.remission
##   <fct>               <int>         <int>
## 1 6-MP                    6            35
## 2 control                 1            23
```

1.8.1 Summarise Across

Use built-in dataset survey, showing student data. Obtain the average for each numeric column. Note the use of "across" for all numeric columns in this syntax:

```
library(MASS)
```

Get the first ten rows and copy to a new dataframe:

```
subset.survey <- survey[1:10,]
library(dplyr)
head(subset.survey)
```

```
##        Sex Wr.Hnd NW.Hnd W.Hnd     Fold Pulse    Clap Exer Smoke
Height     M.I
## 1 Female   18.5   18.0 Right  R on L     92    Left Some Never
173.00    Metric
## 2   Male   19.5   20.5 Left   R on L    104    Left None Regul 177.80
Imperial
## 3   Male   18.0   13.3 Right  L on R     87 Neither None
Occas      NA    <NA>
## 4   Male   18.8   18.9 Right  R on L     NA Neither None Never
160.00    Metric
## 5   Male   20.0   20.0 Right Neither     35   Right Some Never
165.00    Metric
## 6 Female   18.0   17.7 Right  L on R     64   Right Some Never 172.72
Imperial
##      Age
## 1 18.250
## 2 17.583
## 3 16.917
```

40

```
## 4 20.333
## 5 23.667
## 6 21.000
```

As a slight timesaver, note that head(), with no dataframe specified, uses the dataframe or tibble in the pipe:

```
subset.survey %>%
    na.omit() %>%   #remove any NAs
    group_by(Sex) %>%
    summarise(across(where(is.numeric), mean,
    .names = "mean_{col}")) %>%
    head()
```

```
## `summarise()` ungrouping output (override with `.groups` argument)
## # A tibble: 2 x 6
##    Sex     mean_Wr.Hnd mean_NW.Hnd mean_Pulse mean_Height mean_Age
##    <fct>        <dbl>       <dbl>      <dbl>       <dbl>    <dbl>
## 1 Female        17.8        17.7       76.7        168.     25.0
## 2 Male          19.1        19.2       76.8        174.     20.3
```

msleep is a built-in dataset from the ggplot2 package:

```
new.sleep <- msleep %>%
  group_by(vore, order)
```

Use summarise to count vore order combinations:

```
s <- summarise(new.sleep, n())
```

```
## `summarise()` regrouping output by 'vore' (override with `.groups`
argument)
```

```
s
```

```
## # A tibble: 32 x 3
## # Groups:    vore [5]
##    vore   order            `n()`
##    <chr> <chr>           <int>
## 1 carni Carnivora          12
```

```
##  2 carni Cetacea              3
##  3 carni Cingulata            1
##  4 carni Didelphimorphia      1
##  5 carni Primates             1
##  6 carni Rodentia             1
##  7 herbi Artiodactyla         5
##  8 herbi Diprotodontia        1
##  9 herbi Hyracoidea           2
## 10 herbi Lagomorpha           1
## # ... with 22 more rows
```

If you are just interested in the totals, extend the use of the pipe:

```
new.sleep.totals <- msleep %>%
  group_by(vore, order) %>%
  summarise(n())
```

```
## `summarise()` regrouping output by 'vore' (override with `.groups`
argument)
```

```
new.sleep.totals
```

```
## # A tibble: 32 x 3
## # Groups:   vore [5]
##     vore  order              `n()`
##     <chr> <chr>              <int>
##  1 carni Carnivora            12
##  2 carni Cetacea               3
##  3 carni Cingulata             1
##  4 carni Didelphimorphia       1
##  5 carni Primates              1
##  6 carni Rodentia              1
##  7 herbi Artiodactyla          5
##  8 herbi Diprotodontia         1
##  9 herbi Hyracoidea            2
## 10 herbi Lagomorpha            1
## # ... with 22 more rows
```

1.9 Gathering: Convert Multiple Columns into One

Gather converts multiple columns into one column. To illustrate, the following example concentrates three date-related columns into one. The teen birth rates, per 1000 women, are from www.cdc.gov:

```
state <- c("Maryland", "Alaska", "New Jersey")
income <- c(76067,74444,73702)
median.us <- c(61372,61372,61372)
life.expectancy <- c(78.8,78.3,80.3)
teen.birth.rate.2015 <- c(17,29.3,12.1)
teen.birth.rate.2007 <- c(34.3,42.9,24.9)
teen.birth.rate.1991 <- c(54.1, 66, 41.3)
top.3.states <- data.frame(state, income, median.us,
  life.expectancy,
  teen.birth.rate.2015, teen.birth.rate.2007,
  teen.birth.rate.1991)
names(top.3.states) <- c("state", "income", "median.us",
   "life.expectancy","2015","2007","1991")
top.3.states
```

```
##         state income median.us life.expectancy 2015 2007 1991
## 1    Maryland  76067     61372            78.8 17.0 34.3 54.1
## 2      Alaska  74444     61372            78.3 29.3 42.9 66.0
## 3 New Jersey  73702     61372            80.3 12.1 24.9 41.3
```

Now use gather to put all three years in one column:

```
new.top.3.states <- top.3.states %>%
  gather("2015", "2007", "1991", key = "year", value = "cases")
new.top.3.states
```

```
##         state income median.us life.expectancy year cases
## 1    Maryland  76067     61372            78.8 2015  17.0
## 2      Alaska  74444     61372            78.3 2015  29.3
## 3 New Jersey  73702     61372            80.3 2015  12.1
## 4    Maryland  76067     61372            78.8 2007  34.3
## 5      Alaska  74444     61372            78.3 2007  42.9
```

```
## 6 New Jersey  73702        61372           80.3 2007  24.9
## 7   Maryland  76067        61372           78.8 1991  54.1
## 8     Alaska  74444        61372           78.3 1991  66.0
## 9 New Jersey  73702        61372           80.3 1991  41.3
```

1.10 Spreading: Consolidation of Multiple Rows into One

Sometimes data for the same observation is contained in multiple rows. Spread condenses that information into one row, usually making subsequent calculations far simpler. The following example was obtained from stack overflow.[5] Stack overflow is an excellent site for R help. I'm always impressed by the quality and depth of answers found there.

```
df_1 <- data_frame(Type = c("TypeA", "TypeA", "TypeB", "TypeB"),
  Answer = c("Yes", "No", NA, "No"), n = 1:4)

df_1 #before

## # A tibble: 4 x 3
##    Type  Answer     n
##    <chr> <chr>  <int>
## 1 TypeA Yes        1
## 2 TypeA No         2
## 3 TypeB <NA>       3
## 4 TypeB No         4
df_2 <- df_1 %>%
filter(!is.na(Answer)) %>%
    spread(key=Answer, value=n)
```

---

[5]https://stackoverflow.com/questions/34684925/how-to-use-the-spread-function-properly-in-tidyr. Accessed on January 20, 2021.

After using spread, "No" and "Yes" answers are in separate columns, making summaries and other analysis easier:

```
df_2 #after

## # A tibble: 2 x 3
##   Type    No   Yes
##   <chr> <int> <int>
## 1 TypeA    2     1
## 2 TypeB    4    NA
#
```

1.11 Separate: Divide a Single Column into Multiple Columns

The separate function divides a single column into two or more columns:

```
state <- c("Maryland", "Alaska", "New Jersey")
income <- c(76067,74444,73702)
median.us <- c(61372,61372,61372)
life.expectancy <- c(78.8,78.3,80.3)
teen.birth <- c("17//34.3//54.1", "29.0//42.9//66.0", "12.1//24.9//41.3")
```

teen.birth as a column has three data elements per row, separated by a special character ("//"). Before executing the separate function, the years 2015, 2007, and 1991 are combined in the teen.birth column.

```
top.3.states <- data.frame(state, income, median.us,
  life.expectancy,teen.birth)
top.3.states

##         state income median.us life.expectancy       teen.birth
## 1    Maryland  76067     61372            78.8   17//34.3//54.1
## 2      Alaska  74444     61372            78.3 29.0//42.9//66.0
## 3  New Jersey  73702     61372            80.3 12.1//24.9//41.3
```

Using the separate function, the three years' data crammed into one column are separated out into a far more convenient structure of three columns:

```
top.3.states.separated.years <- top.3.states %>%
    separate(teen.birth,
    into = c("2015", "2007","1991"), sep = "//")
top.3.states.separated.years
```

```
##          state income median.us life.expectancy 2015 2007 1991
## 1    Maryland  76067     61372           78.8   17 34.3 54.1
## 2      Alaska  74444     61372           78.3 29.0 42.9 66.0
## 3 New Jersey  73702     61372           80.3 12.1 24.9 41.3
```

1.12 Recap of Handy DPLYR Functions

DPLYR contains a cornucopia of useful functions to transform your data into what you need to solve analytical problems. Real data is often riddled with missing values, numerically valid but wrong values, oddball structures, and many other variations. Particularly when working with unstructured data, you'll be greatly helped by a solid knowledge of DPLYR, Stringr, Lubridate, and regular expressions (often shortened to RegEx).

The following is a useful toolkit which can be used along with the five key DPLYR functions or, in some cases, stand-alone. The DPLYR manual on CRAN has even more depth (https://cran.r-project.org/web/packages/dplyr/dplyr.pdf) than I show here.

1.12.1 Number of Observations (n) Used Across Multiple DPLYR Functions

A simple group count is a tool for everyday use. The n function applies across mutate, summarise, and filter. The following code, adapted from the DPLYR package documentation (as of November 10, 2018), combines several group count uses in one pipe command.

1.12.2 Basic Counts

```
m <- mutate(new.sleep, kount = n()) #new variable, kount added to extreme right

m[1:5,c(1:4,10:12)] #limit number of columns to fit on page
```

```
## # A tibble: 5 x 7
## # Groups:   vore, order [5]
##   name                       genus      vore  order       brainwt bodywt kount
##   <chr>                      <chr>      <chr> <chr>        <dbl>   <dbl>  <int>
## 1 Cheetah                    Acinonyx   carni Carnivora    NA      50     12
## 2 Owl monkey                 Aotus      omni  Primates     0.0155  0.48   10
## 3 Mountain beaver            Aplodontia herbi Rodentia     NA      1.35   16
## 4 Greater short-tailed shrew Blarina    omni  Soricomorp~  0.00029 0.019   3
## 5 Cow                        Bos        herbi Artiodacty~  0.423   600     5
```

Filter by count of vore order which exceeds 14:

```
f <- filter(new.sleep, n() > 14)
f[1:5,c(1:4,10:11)]
```

```
## # A tibble: 5 x 6
## # Groups:   vore, order [1]
##   name                     genus      vore  order     brainwt bodywt
##   <chr>                    <chr>      <chr> <chr>        <dbl>  <dbl>
## 1 Mountain beaver          Aplodontia herbi Rodentia    NA      1.35
## 2 Guinea pig               Cavis      herbi Rodentia    0.0055  0.728
## 3 Chinchilla               Chinchilla herbi Rodentia    0.0064  0.42
## 4 Western american chipmunk Eutamias  herbi Rodentia    NA      0.071
## 5 Mongolian gerbil         Meriones   herbi Rodentia    NA      0.053
```

1.12.3 Nth Functions

First entry:

```
salary.description <- c("Golden parachute type","Well to do",
"Average","Below average", "bring date seeds instead of flowers")
first(salary.description)
```

```
## [1] "Golden parachute type"
```

Last entry:

```
last(salary.description)
```

```
## [1] "bring date seeds instead of flowers"
```

Third from the end:

```
nth(salary.description, -3)
```

```
## [1] "Average"
```

Second element of vector:

```
nth(salary.description,2)
```

```
## [1] "Well to do"
```

1.12.4 Count Distinct Values

It is handy to determine how many unique values exist in a vector. Use n_distinct() as follows.

Create a nine-element vector, some of which are not unique:

```
a.vector <- c(22,33,44,1,2,3,3,3,4)
original.length <- length(a.vector)
original.length
```

```
## [1] 9
```

Show number of distinct elements only (7):

```
distinct.a.vector <- n_distinct(a.vector)
distinct.a.vector
```

```
## [1] 7
```

```
test1 <- if_else(original.length == distinct.a.vector, "all values
unique","some duplicate values in vector")
test1
```

```
## [1] "some duplicate values in vector"
```

Now try with both vectors containing unique values:

```
b.vector <- c(1,2,3,4,5,6)
length(b.vector)
```

```
## [1] 6
```

```
distinct.b.vector <- n_distinct(b.vector)
distinct.b.vector #show count (length) of distinct numbers
```

```
## [1] 6
```

```
test2 <- if_else(length(b.vector) == distinct.b.vector, "all values
unique", "duplicates")
test2
```

```
## [1] "all values unique"
```

1.12.5 na_if

You can set a calculation value to a specified value if NA is calculated. For example, division by zero will result in "inf" in the following calculation:

```
test <- c(100, 0, 999)
x <- 5000/test
x
```

```
## [1] 50.000000       Inf  5.005005
```

Rather than having a value of infinite in the resulting vector, we can tell R to substitute NA rather than Inf:

```
x <- 5000/na_if(test,0) # if any zero occurs in test,
x
```

```
## [1] 50.000000        NA  5.005005
```

```
class(x) #use class to show the type of variable
```

```
## [1] "numeric"
```

1.12.6 Coalesce to Replace Missing Values

Replace missing values with zero or some other value using the coalesce function:

```
x <- c(33,4,11,NA,9)
x
```

```
## [1] 33  4 11 NA  9
```

```
x <- coalesce(x,0)
x
```

```
## [1] 33  4 11  0  9
```

1.13 Ranking Functions

Values in a vector can be ranked by a variety of methods. The following are some of DPLYR's variations.

1.13.1 Ranking via Index

This example shows index numbers of the vector corresponding to each elements. Ranking shows the 6 as the first element, which means that the 6th element of y, which equals 3, has the lowest value, ascending. The index value of 1, which occurs on the extreme right of rank1, corresponds to the highest value in the vector, 100.

```
y <- c(100,4,12,6,8,3)
rank1 <-row_number(y)
rank1
```

```
## [1] 6 2 5 3 4 1
y[rank1[1]] #lowest rank; in this case, rank1[1] points to # y[6] which is
3 (lowest)
```

```
## [1] 3
y[rank1[6]] #highest ranking number, in this case the first # number, 100
## [1] 100
```

1.13.2 Minimum Rank

```
rank2 <- min_rank(y) #in this specific case (for y), gives same results as
#row_number
rank2
```

```
## [1] 6 2 5 3 4 1
```

1.13.3 Dense Rank[6]

```
rank3 <- dense_rank(y)
rank3
```

```
## [1] 6 2 5 3 4 1
```

1.13.4 Percent Rank

The first element of y is in the 100 percentile; the second element of y is in the 2 percentile; the last element of y is in the 0 percentile:

```
rank4 <- percent_rank(y)
rank4
```

```
## [1] 1.0 0.2 0.8 0.4 0.6 0.0
```

1.13.5 Cumulative Distribution Function

This function shows the proportion of all values less than or equal to the current rank:

```
y <- c(100,4,12,6,8,3)
rank5 <- cume_dist(y)
rank5
```

---

[6]"Window function: returns the rank of rows within a window partition, without any gaps. The difference between rank and dense_rank is that dense_rank leaves no gaps in ranking sequence when there are ties. That is, if you were ranking a competition using dense_rank and had three people tie for second place, you would say that all three were in second place and that the next person came in third." "R: Dense_rank," accessed on January 20, 2021, https://spark.apache.org/docs/1.6.1/api/R/dense_rank.html.

```
## [1] 1.0000000 0.3333333 0.8333333 0.5000000 0.6666667 0.1666667
```

Break the input vector into n buckets:

```
rank6 = ntile(y, 3) #in this case, choose 3 buckets
rank6
```

```
## [1] 3 1 3 2 2 1
```

The base R quantile function also has an easy-to-read output:

```
test.vector <- c(2,22,33,44,77,89,99)
quantile(test.vector, prob = seq(0,1,length = 11),type = 5)
```

```
##    0%   10%   20%   30%   40%   50%   60%   70%   80%   90% 100%
##   2.0   6.0  20.0  28.6  36.3  44.0  67.1  81.8  90.0  97.0  99.0
```

1.14 Sampling

Sampling serves many purposes. In some cases, you may simply lack the time or computer resources to process massive datasets. Using a standard but high-powered laptop, running through 50–250 million records is feasible, particularly if you vectorize functions rather than using programming loops. The use of tibbles rather than dataframes is another way to speed up processing. But at some point, very large datasets become unwieldy, and either sampling or intermediate summarization will be required (unless you have powerful resources available). Sampling can often provide an accurate data profile and allow for reasonably accurate visualizations. It is also handy when developing your analytics system—don't debug your base logic by dragging around 50M records. Sample first.

Examples:

Randomly sample 5 rows out of ChickWeight's 578 entries:

```
data("ChickWeight")
my.sample <- sample_n(ChickWeight, 5)
my.sample
```

```
##    weight Time Chick Diet
## 1      42    0     1    1
## 2      66    4    41    4
```

```
## 3      45    2    16    1
## 4     128   10    14    1
## 5      86    6    21    2
```

Set a seed number each time you want a sample which can be reproduced by others. Otherwise, each time the routine is run, different results may be obtained:

```
set.seed(833)
```

Sampling with replacement = TRUE means that you could get the same row or element more than once. Your choice of True or False for replacement depends on your purpose. If, for example, you are investigating manufacturing defects, you might want to use replace = FALSE since you don't want to waste your time investigating the same defect again:

```
my.sample <- sample_n(ChickWeight, 10, replace = TRUE)
my.sample
```

```
##     weight Time Chick Diet
## 1       98    8    45    4
## 2       42    0    17    1
## 3       98    8    36    3
## 4       51    2    11    1
## 5      198   20     3    1
## 6      237   21    49    4
## 7      205   16    50    4
## 8      170   16    39    3
## 9      332   18    35    3
## 10     144   14    33    3
```

In some cases, sampling needs to be biased toward some higher-impact data element. For example, if you are verifying the accuracy of invoices, you may want to weight large dollar amounts more than smaller amounts. As a result, you are more likely to get a high-value invoice than one with a low dollar amount. In auditing, this practice is sometimes called "dollar unit sampling."

In this example, cars with more cylinders are more likely to be selected as part of the sample:

```
my.sample <- sample_n(mtcars, 12, weight = cyl)
my.sample[,1:5]
```

```
##                      mpg cyl  disp  hp drat
## AMC Javelin         15.2   8 304.0 150 3.15
## Porsche 914-2       26.0   4 120.3  91 4.43
## Merc 280            19.2   6 167.6 123 3.92
## Cadillac Fleetwood  10.4   8 472.0 205 2.93
## Merc 240D           24.4   4 146.7  62 3.69
## Datsun 710          22.8   4 108.0  93 3.85
## Merc 280C           17.8   6 167.6 123 3.92
## Mazda RX4 Wag       21.0   6 160.0 110 3.90
## Merc 450SLC         15.2   8 275.8 180 3.07
## Chrysler Imperial   14.7   8 440.0 230 3.23
## Maserati Bora       15.0   8 301.0 335 3.54
## Valiant             18.1   6 225.0 105 2.76
```

Use sample_frac to obtain a sample equal to a specific percentage of the dataframe rows:

```
test1 <- sample_frac(ChickWeight, 0.02)
test1
```

```
##    weight Time Chick Diet
## 1      48    2    13    1
## 2      62    6    12    1
## 3     197   20    45    4
## 4     234   18    42    4
## 5      58    4    28    2
## 6     163   16     3    1
## 7     103    8    41    4
## 8     103    8    42    4
## 9     120   18    19    1
## 10     48    2    36    3
## 11     80    6    48    4
## 12    137   12    33    3
```

In this example, group_by identifies starwars characters by hair group, and then 7% of the records in each group is selected. This is handy when you want a set percentage from groups whose sizes vary.

```
by_hair_color <- starwars %>% group_by(hair_color)
my.sample <- sample_frac(by_hair_color, .07, replace = TRUE)

#sample 7% with replacement

my.sample[,1:5]

## # A tibble: 5 x 5
## # Groups:   hair_color [3]
##    name        height  mass hair_color skin_color
##    <chr>        <int> <dbl> <chr>      <chr>
## 1 Eeth Koth      171    NA black       brown
## 2 Dormé          165    NA brown       light
## 3 Sebulba        112    40 none        grey, red
## 4 Shaak Ti       178    57 none        red, blue, white
## 5 Tion Medon     206    80 none        grey
```

Tally and count provide basic counts and counts by group.

```
row.kount.only <- ChickWeight %>% tally()
row.kount.only

##n
##1 578

diet.kount <- ChickWeight %>% count(Diet)

diet.kount

##   Diet   n
## 1    1 220
## 2    2 120
## 3    3 120
## 4    4 118
```

1.15 Miscellaneous DPLYR Functions

1.15.1 add_count for Groupwise Filtering

This starwars example is adapted from the official DPLYR documentation.[7] Here, only species that have a single member are shown. add_count is useful for groupwise filtering.

   DPLYR does not provide you functionality that could not possibly be done with base R, Python, and so on. Its great value is efficiency, clarity, and time savings.

```
single.species.kount <- starwars %>%
  add_count(species) %>%
  filter(n == 1)
single.species.kount[,1:6]
```

```
## # A tibble: 29 x 6
##     name                 height  mass hair_color skin_color        eye_color
##     <chr>                 <int> <dbl> <chr>      <chr>             <chr>
##  1 Greedo                  173    74 <NA>       green             black
##  2 Jabba Desilijic Tiure   175  1358 <NA>       green-tan, brown  orange
##  3 Yoda                     66    17 white      green             brown
##  4 Bossk                   190   113 none       green             red
##  5 Ackbar                  180    83 none       brown mottle      orange
##  6 Wicket Systri Warrick    88    20 brown      brown             brown
##  7 Nien Nunb               160    68 none       grey              black
##  8 Nute Gunray             191    90 none       mottled green     red
##  9 Watto                   137    NA black      blue, grey        yellow
## 10 Sebulba                 112    40 none       grey, red         orange
## # ... with 19 more rows
```

1.15.2 Rename

```
mtcars <- rename(mtcars, spam_mpg = mpg)

data(mtcars)
names(mtcars)
```

---

[7]https://cran.r-project.org/web/packages/dplyr/dplyr.pdf. Accessed on January 20, 2021.

```
## [1] "mpg"  "cyl"  "disp" "hp"  "drat" "wt" "qsec" "vs" "am"  "gear"
## [11] "carb"

mtcars <- rename(mtcars, spam_mpg = mpg)
names(mtcars)

## [1] "spam_mpg" "cyl"     "disp"     "hp"       "drat"      "wt"
## [7] "qsec"        "vs"      "am"       "gear"     "carb"
```

1.15.3 case_when

case_when is particularly useful inside mutate. You can create a new variable that relies on a complex combination of existing variables.[8]

```
data(starwars)

new.starwars <- starwars %>%
  dplyr::select(name, mass, gender, species, height) %>%
  mutate(type = case_when(height > 200 | mass > 200 ~ "large",
  species == "Droid" ~ "robot", TRUE ~ "other"))
new.starwars

## # A tibble: 87 x 6
##    name               mass gender    species height type
##    <chr>             <dbl> <chr>     <chr>    <int> <chr>
##  1 Luke Skywalker       77 masculine Human      172 other
##  2 C-3PO                75 masculine Droid      167 robot
##  3 R2-D2                32 masculine Droid       96 robot
##  4 Darth Vader         136 masculine Human      202 large
##  5 Leia Organa          49 feminine  Human      150 other
##  6 Owen Lars           120 masculine Human      178 other
##  7 Beru Whitesun lars   75 feminine  Human      165 other
##  8 R5-D4                32 masculine Droid       97 robot
##  9 Biggs Darklighter    84 masculine Human      183 other
## 10 Obi-Wan Kenobi       77 masculine Human      182 other
## # ... with 77 more rows
```

[8]https://dplyr.tidyverse.org/reference/case_when.html. Accessed on January 20, 2021.

Important note   Sometimes R packages have functions with the same name that causes endless confusion. Both MASS and DPLYR have the same function, "select". To ensure that the DPLYR package is used, use **dplyr::select** as the function. That syntax forces R to use the DPLYR rather than MASS version of select.

CHAPTER 2

Stringr

2.1 Introduction

The moving finger writes; and, having writ, moves on: nor all thy piety nor wit shall lure it back to cancel half a line, nor all thy tears wash out a word of it.

—Rubaiyat of Omar Khayyam [time is unidirectional]

The next two packages, Lubridate and Stringr, omit many exceptions and tricky, oddball situations that standard manuals include by necessity. From a technical perspective, there is nothing new in this book. Indeed, much of the code is copied, with slight modifications, from the excellent, free online manuals. The narrow purpose of this book is to give you enough knowledge to use the packages as quickly as possible. If you have programming experience, the explanations and examples here should have you competent within a day or two.

The results of code execution are shown with "##" by R output. The package knitr formatted the text. In topics I felt were a little abstract for the beginning analyst, I either wrote my own code or searched the Internet for simpler examples. The package manuals do not typically show the result of the code they include. This may be to encourage readers to run code themselves. Back in the day, colleges had a term *in loco parentis*, meaning that the college was a substitute parent. Essentially the idea was to make students do what was good for them, whether eating broccoli or working through problems line by line. Here, I have assumed readers are adults with many demands on their time and attention. Choose the topics of interest and copy to get started. This is a "gateway" book.

© William Yarberry 2021
W. Yarberry, *CRAN Recipes*, https://doi.org/10.1007/978-1-4842-6876-6_2

None of this is to say the manuals and many excellent R books are unnecessary. They simply have a different priority—depth rather than learning speed.

The Stringr package works with character data. For those who start their day with black coffee and a hair shirt and then run up a mountain with their accountant on their back, perhaps memorizing traditional regular expressions works. For all others, Stringr is a wonderfully simple and consistent package. It is intuitive and adaptive and covers 99% of what most people need to do with character data. Regular expressions remind me of one of my first programming jobs, writing in mainframe assembly. Nothing could execute faster or have less ambiguity. And almost nothing could be more tedious to write and maintain by others. With Stringr, it is usually clear what you are doing. Having bad-mouthed regular expressions (often termed *RegEx*), I nonetheless include an entire chapter later in the book devoted to their use. It is worth getting a sense of what these powerful tools can do, should you need them for complex matching tasks.

Next is Lubridate, a package which simplifies date/time processes relative to base R. The entire topic is like the game of Go. It seems simple only if you haven't played. Date/times are complex. A second is a second, but an hour or a day or a year will vary based on when in time it occurs. Lubridate handles just about every date-time calculation anyone can envision and then some. Learning this package will save time compared to all the spot solutions in base R. If nothing else, nail down conversion of an imported Excel/CSV character date to a true R date, which then opens a vast toolbox of date/time analytics.

2.2 Stringr Functions

2.2.1 Find, Count, and Extract

Find where a specific string is located in a larger string; show row and character position:

```
bk.fruit <- c("apQple", "banana", "pear", "pinQeapple")
bk.where.is.Q <-  str_locate(bk.fruit,"Q")
bk.where.is.Q
```

```
##      start end
## [1,]    3    3
## [2,]   NA   NA
## [3,]   NA   NA
## [4,]    4    4
```

The first element of the string vector has a Q in the third position. Hence, str_locate shows row one as having both start and end in position 3 for that first row apQple. The next Q is in the fourth element, pinQeapple, and starts and ends in position 4 for that fourth row.

Next, insert a Q at the end of banana. It shows as seventh character of the second row:

```
bk.fruit <- c("apQple", "bananaQ", "pear", "pinQeapple")
bk.where.is.Q <-  str_locate(bk.fruit,"Q")
bk.where.is.Q
```

```
##       start end
## [1,]     3   3
## [2,]     7   7
## [3,]    NA  NA
## [4,]     4   4
```

```
class(bk.where.is.Q)
```

```
## [1] "matrix" "array"
```

Note that the output is a matrix.

2.2.2 String Detect

Modifiers control matching behavior with modifier functions. It gives instructions to Stringr, specifying how to match up strings/characters. Do you want to ignore case or just use beginning or end characters? Stringr is flexible, consistent, and more intuitive than legacy string functions.

```
bk.pattern <- "a.b"
bk.strings <- c("abb", "a.b")
bk.detect.strings <- str_detect(bk.strings, bk.pattern)
bk.detect.strings
```

```
## [1] TRUE TRUE
```

The pattern or structure was a followed by b. It was found true in both cases:

```
bk.detect.strings.fixed <- str_detect(bk.strings, fixed(bk.pattern))
bk.detect.strings.fixed
```

```
## [1] FALSE   TRUE
```

Literal matching, so abb does not match the a.b pattern from the ICU users guide: `http://userguide.icu-project.org/collation/customization`.

Collation Rule. A Rule-Based Collator is built from a rule string which changes the sort order of some characters and strings relative to the default order. A tailoring is specified via a string containing a set of rules.

Compare strings using standard collation rules:

```
bk.detect.strings.collation.rules <- str_detect(bk.strings, coll(bk.pattern))

bk.detect.strings.collation.rules
```

```
## [1] FALSE   TRUE
```

Some of the Stringr functions are cognizant of locale (country/region where different capitalization and other rules may apply). Locale = "en" (English) is the default. Appendix E in the book lists locales and their codes for most locations. You can also find the information at bit.ly/IS0639-l (ISO639-1 is the applicable standard).

coll() is useful for locale-aware case-insensitive matching. "locale-aware" is an adult version of "Where in the World Is Carmen Sandiego?" Sherlock alert: The first time I saw "\u0130", it did not strike me as an obvious code. It is actually a code for a Latin i—it is an i but has a small horizontal line near the top. (See the following second element.)

```
bk.i <- c("I", "\u0130", "i")
bk.i
```

```
## [1] "I" "I" "i"
```

```
bk.detect.fixed.i <- str_detect(bk.i, fixed("i", TRUE))
bk.detect.fixed.i
```

```
## [1]  TRUE FALSE   TRUE
```

When no locale is specified, "en" (English) is implied:

```
bk.detect.collation <- str_detect(bk.i, coll("i", TRUE))
bk.detect.collation
```

```
## [1]  TRUE FALSE  TRUE
```

Now specify the location as Turkey:

```
bk.detect.using.turkey <- str_detect(bk.i, coll("i", TRUE, locale = "tr"))
bk.detect.using.turkey
```

```
## [1] FALSE  TRUE  TRUE
```

2.2.3 String Count

A different answer is obtained when "tr" is substituted for the default "en". Count number of matches in a string:

```
bk.fruit <- c("apple", "banana", "pear", "pineapple")
bk.fruit.count <- str_count(bk.fruit, "a")
bk.fruit.count
```

```
## [1] 1 3 1 1
```

```
bk.count.with.p <-str_count(bk.fruit, "p")
bk.count.with.p
```

```
## [1] 2 0 1 3
```

```
bk.count.with.e <- str_count(bk.fruit, "e")
bk.count.with.e
```

```
## [1] 1 0 1 2
```

```
bk.count.multiple <- str_count(bk.fruit, c("a", "b", "p", "p"))
bk.count.multiple
```

```
## [1] 1 1 1 3
```

```
bk.count <- str_count(c("a.", "...", ".a.a"), ".")
bk.count
## [1] 2 3 4
```

Match fixed string:

```
bk.fixed <- str_count(c("a.", "...", ".a.a"), fixed("."))
bk.fixed
```

```
## [1] 1 3 2
```

Number of periods in each string. Detect the existence of a pattern in a string:

```
fruit <- c("apple", "banana", "pear", "pineapple")
str_detect(fruit, "a")
```

```
## [1] TRUE TRUE TRUE TRUE
```

```
str_detect(fruit, "^a")
```

```
## [1]  TRUE FALSE FALSE FALSE
```

```
str_detect(fruit, "a$")
```

```
## [1] FALSE  TRUE FALSE FALSE
```

```
str_detect(fruit, "b")
```

```
## [1] FALSE  TRUE FALSE FALSE
```

```
str_detect(fruit, "[aeiou]")
```

```
## [1] TRUE TRUE TRUE TRUE
```

Also vectorized over pattern:

```
str_detect("aecfg", letters)
```

```
##  [1]  TRUE FALSE  TRUE FALSE  TRUE  TRUE  TRUE FALSE FALSE FALSE FALSE
     FALSE
## [13] FALSE FALSE FALSE FALSE FALSE FALSE FALSE FALSE FALSE FALSE FALSE
     FALSE
## [25] FALSE FALSE
```

Returns TRUE if the pattern does NOT match:

```
str_detect(fruit, "^p", negate = TRUE)
```

```
## [1]  TRUE  TRUE FALSE FALSE
```

Duplicate or concatenate strings within a string:

```
fruit <- c("apple", "pear", "banana")
str_dup(fruit, 2)
```

```
## [1] "appleapple"    "pearpear"       "bananabanana"
str_dup(fruit, 1:3)
```

```
## [1] "apple"                "pearpear"                "bananabananabanana"
```

Find "na" and then add on multiples of it:

```
str_c("ba", str_dup("na", 0:5))
```

```
## [1] "ba"            "bana"         "banana"        "bananana"       "banananana"
## [6] "bananananana"
```

2.2.4 String Remove

```
fruits <- c("one apple", "two pears", "three bananas")
str_remove(fruits, "[aeiou]")
```

```
## [1] "ne apple"       "tw pears"       "thre bananas"
```

To relieve the tedium of using the same old fruit examples, I'm using part of a speech from a funeral oration by Pericles in the 5th-century BCE Athens:[1] bk.speech is a single string containing an entire sentence. bk.speech.words is a character vector containing multiple words from the same sentence.

```
bk.speech <- "While we are thus unconstrained in our private business, a
spirit of reverence pervades our public acts"
bk.speech.words <- c("While", "we", "are", "thus", "unconstrained", "in",
"our", "private", "business,",
"a", "spirit", "of", "reverence", "pervades", "our", "public", "acts")
```

---

[1]"Thucydides: Pericles' Funeral Oration," accessed on February 13, 2021, http://hrlibrary.umn.edu/education/thucydides.html.

First instance of the vowel removed[2]:

```
str_remove_all(fruits, "[aeiou]")

## [1] "n ppl"     "tw prs"     "thr bnns"
```

Just to use something other than fruit, some of the following code will use new strings, as follows:

```
bk.speech <- "While we are thus unconstrained in our private business, a
spirit of reverence pervades our public acts"
bk.speech.words <- c("While", "we", "are", "thus", "unconstrained", "in",
"our", "private", "business,",
"a", "spirit", "of", "reverence", "pervades", "our", "public", "acts")
```

All vowels removed:

```
bk.no.vowels <- str_remove_all(bk.speech, "[aeiou]")
bk.no.vowels

## [1] "Whl w r ths ncnstrnd n r prvt bsnss,  sprt f rvrnc prvds r pblc cts"
```

2.2.5 String Replace

Replace vowels in the *first* occurrence with three z's:

```
bk.vowels.replaced <- str_replace(bk.speech, "[aeiou]", "zzz")
bk.vowels.replaced

## [1] "Whzzzle we are thus unconstrained in our private business, a spirit
of reverence pervades our public acts"
```

Replace vowels in all occurrences with three z's:

```
bk.all.vowels.replaced <- str_replace_all(bk.speech, "[aeiou]", "zzz")
bk.all.vowels.replaced
```

---

[2]https://cran.r-project.org/web/packages/stringr/stringr.pdf, p. 19, accessed on February 14, 2021.

```
## [1] "Whzzzlzzz wzzz zzzrzzz thzzzs zzznczzznstrzzzzzzznzzzd zzzn zzzzzzr
przzzvzzztzzz bzzzszzznzzzss, zzz spzzzrzzzt zzzf rzzzvzzzrzzzzznczzz
pzzzrvzzzdzzzs zzzzzzr pzzzblzzzc zzzcts"
```

Capitalize all vowels.

```
bk.all.vowels.capitalized <- str_replace_all(bk.speech, "[aeiou]", toupper)
bk.all.vowels.capitalized
```

```
## [1] "WhIlE wE ArE thUs UncOnstrAInEd In OUr prIvAtE bUsInEss, A spIrIt
Of rEvErEncE pErvAdEs OUr pUblIc Acts"
```

Replace any "in" with "NA" (using multi-element string bk.speech.words). Note how any element containing "in" is replaced, not simply "in" as an exact match:

```
bk.in.replace.with.na <- str_replace_all(bk.speech.words, "in", NA_
character_)
bk.in.replace.with.na
```

```
##  [1] "While"     "we"        "are"       "thus"      NA          NA
##  [7] "our"       "private"   NA          "a"         "spirit"    "of"
## [13] "reverence" "pervades"  "our"       "public"    "acts"
```

In this example, we find the first instance of a vowel and then add to it the same vowel. If you suspect the "\\" construction looks like RegEx syntax, you'd be correct:

```
bk.dup.first.vowel <- str_replace(bk.speech.words,
  "([aeiou])", "\\1\\1")
bk.dup.first.vowel
```

```
##  [1] "Whiile"         "wee"        "aare"      "thuus"
##  [5] "uunconstrained" "iin"        "oour"      "priivate"
##  [9] "buusiness,"     "aa"         "spiirit"   "oof"
## [13] "reeverence"     "peervades"  "oour"      "puublic"
## [17] "aacts"
```

Replace first instance of a vowel with a number. Note how the numbers, starting with 7, are recycled as repeated replacements occur.

```
bk.replace.vowel.w.number <- str_replace(bk.speech.words, "[aeiou]",
c("7", "8", "9"))
```

```
bk.replace.vowel.w.number
```

```
##  [1] "Wh7le"          "w8"          "9re"        "th7s"
##  [5] "8nconstrained" "9n"          "7ur"        "pr8vate"
##  [9] "b9siness,"      "7"           "sp8rit"     "9f"
## [13] "r7verence"      "p8rvades"    "9ur"        "p7blic"
## [17] "8cts"
```

We have been using vowels in these examples, but any text works. When you run this code, you'll get a warning message about object length. "ou" has two characters but is being replaced by a single character digit.

In general, you should stay away from recycling like this, even though recycling is often used in real-world code. To me, frequent reliance on recycling looks like a coding error in the making. Even worse, the code may run and look correct at first glance.

```
bk.replace.ou <- str_replace(bk.speech.words, "ou",
  c("7", "8", "9"))
bk.replace.ou
```

```
##  [1] "While"          "we"          "are"        "thus"
##  [5] "unconstrained" "in"          "7r"         "private"
##  [9] "business,"      "a"           "spirit"     "of"
## [13] "reverence"      "pervades"    "9r"         "public"
## [17] "acts"
```

Use str_c and a pipe coding structure to "collapse" several strings into one AND substitute for some characters at the same time. Note that "thus" is replaced in this logic but "while" is not. str_replace_all is case sensitive.

Insert four dollar signs via the collapse command, between string elements:

```
bk.make.one.altered.string <- bk.speech.words %>%
  str_c(collapse = "$$$$") %>%
  str_replace_all(c("while" = "____WHILE____",
  "thus" = "____THEREFORE____"))
bk.make.one.altered.string
```

```
## [1] "While$$$$we$$$$are$$$$____THEREFORE____$$$$unconstrained$$$$in$$$$
our$$$$private$$$$business,$$$$a$$$$spirit$$$$of$$$$reverence$$$$pervades$$
$$our$$$$public$$$$acts"
```

2.2.6 String Starts

Use str_starts to identify the presence or absence of a character(s).

```
bk.starts.with.p <- str_starts(bk.speech.words, "p")
bk.starts.with.p  #logical output
```

```
##  [1] FALSE FALSE FALSE FALSE FALSE FALSE FALSE  TRUE FALSE FALSE FALSE
     FALSE
## [13] FALSE  TRUE FALSE  TRUE FALSE
```

Using negate, show the absence of a starting character(s).

```
bk.starts.with.no.p <- str_starts(bk.speech.words,
  "p", negate = TRUE)
bk.starts.with.no.p
```

```
##  [1]  TRUE  TRUE  TRUE  TRUE  TRUE  TRUE  TRUE
FALSE  TRUE  TRUE  TRUE  TRUE
## [13]  TRUE FALSE  TRUE FALSE  TRUE
```

2.2.7 String Ends

To work with ending characters, use str_ends.

```
bk.ends.with <- str_ends(bk.speech.words, "s")
bk.ends.with
```

```
##  [1] FALSE FALSE FALSE  TRUE FALSE FALSE FALSE FALSE FALSE FALSE FALSE
     FALSE
## [13] FALSE  TRUE FALSE FALSE  TRUE
```

Using negate, show the absence of an ending character(s).

```
bk.ends.with.no.s <- str_ends(bk.speech.words, "s",
  negate = TRUE)
bk.ends.with.no.s
```

```
##  [1]  TRUE  TRUE  TRUE FALSE  TRUE  TRUE  TRUE  TRUE  TRUE  TRUE
     TRUE  TRUE
## [13]  TRUE FALSE  TRUE  TRUE FALSE
```

2.2.8 String Subset

str_subset keeps strings matching a pattern. In this example, only strings with a "t" somewhere in them will be included:

```
bk.subset <- str_subset(bk.speech.words, "t")
bk.subset
```

```
## [1] "thus"          "unconstrained" "private"       "spirit"
## [5] "acts"
```

Return elements which do NOT match. "unconstrained" is missing in the output.

```
bk.subset.no.match <- str_subset(bk.speech.words,
   "un", negate = TRUE)
bk.subset.no.match
```

```
##  [1] "While"    "we"        "are"      "thus"     "in"      "our"
##  [7] "private"  "business," "a"
"spirit"    "of"        "reverence"
## [13] "pervades" "our"       "public"   "acts"
```

Show index of matching elements. Only the 11th element contains characters "sp":

```
bk.which.match <- str_which(bk.speech.words, "sp")
bk.which.match
```

```
## [1] 11
```

Show strings with start with a particular string. As in other regular expressions, ^ is an anchor character used in regular expressions, indicating a starting position. str_subset in this case shows the actual strings, not the index number.

```
bk.starts.with.and.includes.entire.string <-
    str_subset(bk.speech.words,  "^a")
bk.starts.with.and.includes.entire.string
```

```
## [1] "are" "a"     "acts"
```

Note the RegEx use of the dollar sign.

```
bk.ends.with.and.includes.entire.string <-
    str_subset(bk.speech.words, "e$")
bk.ends.with.and.includes.entire.string
```

```
## [1] "While"      "we"         "are"        "private"    "reverence"
```

Key point   Missing data, NA, never matches anything.

A period in quotes means anything, like a wildcard character, with the exception of a newline character.

```
bk.test.na <- str_subset(c("qqq", NA, "mmm", "9999"), ".")
bk.test.na
```

```
## [1] "qqq"  "mmm"  "9999"
```

You will get the same result using str_which, except that str_which gives an index number in the output.

```
bk.test.na <- str_which(c("qqq", NA, "mmm", "9999"), ".")
bk.test.na
```

```
## [1] 1 3 4
```

2.2.9 String Which

The following is a good example of str_which adapted from a University of Michigan statistics course[3]:

```
bk.lorem <- c("Lorem ipsum dolor sit amet, consectetur
 adipiscing elit, sed do eiusmod tempor incididunt ut labore et
 dolore magna aliqua.", "Ut enim ad minim veniam, quis nostrud
 exercitation ullamco laboris nisi ut aliquip ex ea commodo
 consequat.", "Duis aute irure dolor in reprehenderit in
```

---

[3]"String Manipulation," accessed on January 30, 2021, http://dept.stat.lsa.umich. edu/~jerrick/courses/stat701/notes/stringmanip.html.

```
voluptate velit esse cillum dolore eu fugiat nulla pariatur.",
"Excepteur sint occaecat cupidatat non proident, sunt in culpa
qui officia deserunt mollit anim id est laborum.")
```

```
length(bk.lorem)
```

```
## [1] 4
```

There are four elements in this vector:

```
str_which(bk.lorem, "dolor")
```

```
## [1] 1 3
```

The string "dolor" is found in the first element of the vector bk.lorem, third string within the first element.

2.2.10 String Extraction Using Regular Expressions

For further extraction examples, I'll use a sentence from a Wikipedia entry on the rocket pioneer Robert Goddard.[4]

```
bk.goddard.facts <- "He and his team launched 34 rockets between 1926 and
1941, achieving altitudes as high as 2.6 km (1.6 mi) and speeds as fast as
885 km/h (550 mph)."
```

```
bk.goddard.words <- c("He and his team launched ", "34 rockets ", "between
  1926 and 1941", ", achieving altitudes as high as ", "2.6 km (1.6 mi) ",
  "and speeds as fast as ", "885 km/h (550 mph)")
```

Extract numbers using \\d RegEx syntax.

```
bk.goddard.facts.numbers <- str_extract(bk.goddard.facts, "\\d")
bk.goddard.facts.numbers  #first number encountered
```

```
## [1] "3"
```

---

[4]"Robert H. Goddard," in Wikipedia, February 14, 2021, https://en.wikipedia.org/w/index.
php?title=Robert_H._Goddard&oldid=1006681738.

```
bk.goddard.facts.numbers <- str_extract(bk.goddard.facts,
  "\\d+")
bk.goddard.facts.numbers  #entire (all digits) first number encountered
```

```
## [1] "34"
```

Get the lowercase letters, first instance for each element.

```
bk.goddard.all.alphas <- str_extract(bk.goddard.words, "[a-z]")
bk.goddard.all.alphas
```

```
## [1] "e" "r" "b" "a" "k" "a" "k"
```

Get all lowercase letters in the first word of the string.

```
bk.goddard.all.alphas <- str_extract(bk.goddard.words, "[a-z]+")
bk.goddard.all.alphas
```

```
## [1] "e"         "rockets"   "between"   "achieving" "km"         "and"
## [7] "km"
```

Extract all the numbers. Output is a list.

```
z <- gregexpr("[0-9]+",bk.goddard.words)
bk.goddard.all.numbers <- regmatches(bk.goddard.words,z)
bk.goddard.all.numbers
```

```
## [[1]]
## character(0)
##
## [[2]]
## [1] "34"
##
## [[3]]
## [1] "1926" "1941"
##
## [[4]]
## character(0)
##
## [[5]]
```

```
## [1] "2" "6" "1" "6"
##
## [[6]]
## character(0)
##
## [[7]]
## [1] "885" "550"
```

```
class(bk.goddard.all.numbers)
```

```
## [1] "list"
```

Convert to a dataframe (for convenience).

```
bk.goddard.all.numbers <- as.data.frame(matrix(unlist(bk.goddard.all.
numbers), nrow=length(bk.goddard.all.numbers),byrow=TRUE))
```

```
bk.goddard.all.numbers
```

```
##       V1    V2
## 1     34 1926
## 2 1941     2
## 3     6    1
## 4     6   885
## 5   550    34
## 6 1926 1941
## 7    2    6
```

You can also split up the strings into "words" bounded by spaces, using the \\W RegEx syntax.

```
bk.split.into.words <- str_split(bk.goddard.words, "\\W")
bk.split.into.words
```

```
## [[1]]
## [1] "He"       "and"       "his"       "team"       "launched" ""
##
## [[2]]
## [1] "34"        "rockets" ""
```

```
##
## [[3]]
## [1] "between" "1926"      "and"        "1941"
##
## [[4]]
## [1] ""              ""              "achieving" "altitudes" "as"        "high"
## [7] "as"            ""
##
## [[5]]
## [1] "2"  "6"  "km" ""   "1"  "6"  "mi" ""    ""
##
## [[6]]
## [1] "and"     "speeds" "as"       "fast"   "as"       ""
##
## [[7]]
## [1] "885" "km"  "h"   ""     "550" "mph" ""
```

2.2.11 String Extract All

Another approach is to use str_extract_all to extract all words:

```
bk.all.words.extraction <- str_extract_all("Though it be madness, yet there
is method to it", boundary("word"))
bk.all.words.extraction #list
## [[1]]
##  [1] "Though"   "it"       "be"       "madness" "yet"      "there"    "is"
##  [8] "method"   "to"       "it"

shopping_list <- c("apples x4", "bag of flour", "bag of sugar", "milk x2")
str_extract(shopping_list, "\\d")

## [1] "4" NA  NA  "2"

str_extract(shopping_list, "[a-z]+")

## [1] "apples" "bag"      "bag"      "milk"
```

```
str_extract(shopping_list, "[a-z]{1,4}")
```

```
## [1] "apple" "bag"   "bag"   "milk"
```

```
str_extract(shopping_list, "\\b[a-z]{1,4}\\b")
```

```
## [1] NA      "bag"  "bag"   "milk"
```

Extract all matches:

```
str_extract_all(shopping_list, "[a-z]+")
```

```
## [[1]]
## [1] "apples" "x"
##
## [[2]]
## [1] "bag"    "of"     "flour"
##
## [[3]]
## [1] "bag"    "of"     "sugar"
##
## [[4]]
## [1] "milk" "x"
```

```
str_extract_all(shopping_list, "\\b[a-z]+\\b")
```

```
## [[1]]
## [1] "apples"
##
## [[2]]
## [1] "bag"    "of"     "flour"
##
## [[3]]
## [1] "bag"    "of"     "sugar"
##
## [[4]]
## [1] "milk"
```

```
str_extract_all(shopping_list, "\\d")
```

```
## [[1]]
## [1] "4"
##
## [[2]]
## character(0)
##
## [[3]]
## character(0)
##
## [[4]]
## [1] "2"
```

Simplify results into character matrix:

```
str_extract_all(shopping_list, "\\b[a-z]+\\b", simplify = TRUE)
```

```
##        [,1]      [,2] [,3]
## [1,] "apples"  ""    ""
## [2,] "bag"     "of"  "flour"
## [3,] "bag"     "of"  "sugar"
## [4,] "milk"    ""    ""
```

```
str_extract_all(shopping_list, "\\d", simplify = TRUE)
```

```
##        [,1]
## [1,] "4"
## [2,] ""
## [3,] ""
## [4,] "2"
```

Extract all words:

```
str_extract_all("This is, surprisingly, a sentence.", boundary("word"))
```

```
## [[1]]
## [1] "This"         "is"              "surprisingly" "a"              "sentence"
```

Flatten a string—take many strings and combine them into one:

```
letters
```

```
##  [1] "a" "b" "c" "d" "e" "f" "g" "h" "i" "j" "k" "l" "m" "n" "o" "p" "q"
    "r" "s"
## [20] "t" "u" "v" "w" "x" "y" "z"
```

```
str_flatten(letters)
```

```
## [1] "abcdefghijklmnopqrstuvwxyz"
```

```
str_flatten(letters, "-")
```

```
## [1] "a-b-c-d-e-f-g-h-i-j-k-l-m-n-o-p-q-r-s-t-u-v-w-x-y-z"
```

2.2.12 String Glue

Glue is a smart, efficient function. It can be used, among other things, to conveniently intersperse literals with calculated values:

```
name <- "Fred"
age <- 50
anniversary <- as.Date("1991-10-12")
str_glue(
  "My name is {name}, ",
  "my age next year is {age + 1}, ",
  "and my anniversary is {format(anniversary, '%A, %B %d, %Y')}."
)
```

```
## My name is Fred, my age next year is 51, and my anniversary is Saturday,
October 12, 1991.
```

Single braces can be inserted by doubling them:

```
str_glue("My name is {name}, not {{name}}.")
```

```
## My name is Fred, not {name}.
```

```
str_glue(
  "My name is {name}, ",
```

```
  "and my age next year is {age + 1}.",
  name = "Joe",
  age = 40
)
```

```
## My name is Joe, and my age next year is 41.
```

`str_glue_data()` is useful in data pipelines:

```
mtcars %>% str_glue_data("{rownames(.)} has {hp} hp")
```

```
## Mazda RX4 has 110 hp
## Mazda RX4 Wag has 110 hp
## Datsun 710 has 93 hp
## Hornet 4 Drive has 110 hp
## Hornet Sportabout has 175 hp
## Valiant has 105 hp
## Duster 360 has 245 hp
## Merc 240D has 62 hp
## Merc 230 has 95 hp
## Merc 280 has 123 hp
## Merc 280C has 123 hp
## Merc 450SE has 180 hp
## Merc 450SL has 180 hp
## Merc 450SLC has 180 hp
## Cadillac Fleetwood has 205 hp
## Lincoln Continental has 215 hp
## Chrysler Imperial has 230 hp
## Fiat 128 has 66 hp
## Honda Civic has 52 hp
## Toyota Corolla has 65 hp
## Toyota Corona has 97 hp
## Dodge Challenger has 150 hp
## AMC Javelin has 150 hp
## Camaro Z28 has 245 hp
## Pontiac Firebird has 175 hp
## Fiat X1-9 has 66 hp
```

```
## Porsche 914-2 has 91 hp
## Lotus Europa has 113 hp
## Ford Pantera L has 264 hp
## Ferrari Dino has 175 hp
## Maserati Bora has 335 hp
## Volvo 142E has 109 hp
#str_order
```

Order - sort vectors; works like you would expect it to.

2.2.13 String Order (Sorting)

```
str_order(letters)
```

```
##  [1]  1  2  3  4  5  6  7  8  9 10 11 12 13 14 15 16 17 18 19 20 21 22
    23 24 25
## [26] 26
```

"a" is first, "b" is second, and so on:

```
str_sort(letters)
```

```
##  [1] "a" "b" "c" "d" "e" "f" "g" "h" "i" "j" "k" "l" "m" "n" "o" "p" "q"
    "r" "s"
## [20] "t" "u" "v" "w" "x" "y" "z"
```

```
str_order(letters, locale = "haw")
# Locale "Hawaiian (United States)" (haw_US) - LocalePlanet
```

```
##  [1]  1  5  9 15 21  2  3  4  6  7  8 10 11 12 13 14 16 17 18 19 20 22
    23 24 25
## [26] 26
```

```
str_sort(letters, locale = "haw")
```

```
##  [1] "a" "e" "i" "o" "u" "b" "c" "d" "f" "g" "h" "j" "k" "l" "m" "n" "p"
    "q" "r"
## [20] "s" "t" "v" "w" "x" "y" "z"
```

```
bk.x <- c("100a10", "100a5", "2b", "2a")
```

```
bk.x
```

```
## [1] "100a10" "100a5"   "2b"        "2a"
```

```
sort(bk.x)
```

```
## [1] "100a10" "100a5"   "2a"        "2b"
```

```
str_sort(bk.x, numeric = TRUE)
```

```
## [1] "2a"       "2b"        "100a5"   "100a10"
```

2.3 Get or Modify String Information

Word boundaries:

```
bk.words <- c("These are some words.")
bk.boundary.count <- str_count(bk.words, boundary("word"))
bk.boundary.count
```

```
## [1] 4
```

Shows there are four words, as word entities:

```
bk.string.split.blank <- str_split(bk.words, " ")[[1]]
bk.string.split.blank
```

```
## [1] "These"  "are"     "some"    "words."
```

```
class(bk.string.split.blank)
```

```
## [1] "character"
```

The output of the string split function is character:

```
bk.string.split.words <- str_split(bk.words, boundary("word"))[[1]]
bk.string.split.words
```

```
## [1] "These" "are"    "some"   "words"
```

```
length(bk.string.split.words)
```

```
## [1] 4
```

The four words are output as four elements of a vector:

- length of a string
- str_length

```
bk.length <- str_length(letters)
bk.length
```

```
##  [1] 1 1 1 1 1 1 1 1 1 1 1 1 1 1 1 1 1 1 1 1 1 1 1 1 1 1
```

```
length(bk.length)
```

```
## [1] 26
```

```
str_length(NA)
```

```
## [1] NA
```

```
bk.factor.length <- str_length(factor("abc"))
bk.factor.length
```

```
## [1] 3
```

```
str_length(c("i", "like", "programming", NA))
```

```
## [1]  1  4 11 NA
```

There are two ways of representing a u with an umlaut.
Latin small letter u with diaeresis (U+00FC):

```
u1 <- "\u00fc"
```

```
u2 <- stringi::stri_trans_nfd(u1)
```

They print the same

```
u1
```

```
## [1] "ü"
```

```
u2
```

```
## [1] "ü"
```

...but have a different length

```
str_length(u1)
```

```
## [1] 1
```

```
str_length(u2)
```

```
## [1] 2
```

...even though they have the same number of characters:

```
str_count(u1)
```

```
## [1] 1
```

```
str_count(u2)
```

```
## [1] 1
```

Note   The preceding u2 uses the stri_trans_nfd() function. The following is an explanation courtesy: www.rdocumentation.org/packages/stringi/ versions/1.4.3/topics/stri_trans_nfc.

Unicode Normalization Forms are formally defined normalizations of Unicode strings which, for example, make possible to determine whether any two strings are equivalent. Essentially, the Unicode Normalization Algorithm puts all combining marks in a specified order and uses rules for decomposition and composition to transform each string into one of the Unicode Normalization Forms.

The following Normalization Forms (NFs) are supported:

- NFC (Canonical Decomposition, followed by Canonical Composition)

- NFD (Canonical Decomposition)

- NFKC (Compatibility Decomposition, followed by Canonical Composition)

- NFKD (Compatibility Decomposition)

- NFKC_Casefold (combination of NFKC, case folding, and removing ignorable characters which was introduced with Unicode 5.2)

Use str_locate to find the position of patterns in a string.[5]

```
fruit <- c("apple", "banana", "pear", "pineapple")
str_locate(fruit, "$")
```

```
##        start end
## [1,]      6   5
## [2,]      7   6
## [3,]      5   4
## [4,]     10   9
```

In every case in the following, the end is one less than start. This is str_locate's way of telling you there is no match in any of the four elements.

Shows first occurring position of "a" in each character element:

```
str_locate(fruit, "a")
```

```
##        start end
## [1,]      1   1
## [2,]      2   2
## [3,]      3   3
## [4,]      5   5
```

```
str_locate(fruit, "e")
```

```
##        start end
## [1,]      5   5
## [2,]     NA  NA
## [3,]      2   2
## [4,]      4   4
```

```
str_locate(fruit, c("a", "b", "p", "p"))
```

```
##        start end
## [1,]      1   1
## [2,]      1   1
```

---

[5]https://cran.r-project.org/web/packages/stringr/stringr.pdf, p. 15, accessed on February 14, 2021.

```
## [3,]      1    1
## [4,]      1    1
```

```
bk.locate.a <-  str_locate_all(fruit, "a")
bk.locate.a
```

```
## [[1]]
##       start end
## [1,]     1   1
##
## [[2]]
##       start end
## [1,]     2   2
## [2,]     4   4
## [3,]     6   6
##
## [[3]]
##       start end
## [1,]     3   3
##
## [[4]]
##       start end
## [1,]     5   5
```

```
class(bk.locate.a)
```

```
## [1] "list"
```

Note that output is a list:

```
str_locate_all(fruit, "e")
```

```
## [[1]]
##       start end
## [1,]     5   5
##
## [[2]]
##       start end
##
```

```
## [[3]]
##      start end
## [1,]    2   2
##
## [[4]]
##      start end
## [1,]    4   4
## [2,]    9   9
```

```
str_locate_all(fruit, c("a", "b", "p", "p"))
```

```
## [[1]]
##      start end
## [1,]    1   1
##
## [[2]]
##      start end
## [1,]    1   1
##
## [[3]]
##      start end
## [1,]    1   1
##
## [[4]]
##      start end
## [1,]    1   1
## [2,]    6   6
## [3,]    7   7
```

Find location of every character:

```
str_locate_all(fruit, "")
```

```
## [[1]]
##      start end
## [1,]    1   1
## [2,]    2   2
## [3,]    3   3
```

```
## [4,]     4    4
## [5,]     5    5
##
## [[2]]
##       start end
## [1,]     1    1
## [2,]     2    2
## [3,]     3    3
## [4,]     4    4
## [5,]     5    5
## [6,]     6    6
##
## [[3]]
##       start end
## [1,]     1    1
## [2,]     2    2
## [3,]     3    3
## [4,]     4    4
##
## [[4]]
##       start end
##  [1,]     1    1
##  [2,]     2    2
##  [3,]     3    3
##  [4,]     4    4
##  [5,]     5    5
##  [6,]     6    6
##  [7,]     7    7
##  [8,]     8    8
##  [9,]     9    9
```

This includes every character's start/stop location. On list element number 4, nine starts/stops are shown because there are nine characters in the string (pineapple).

Extract matched groups from a string.

This is an efficient way to extract matched groups from a string. If you are, for example, looking for telephone numbers in a blob of unstructured character data, this would save an immense about of data munging time. The following example is from the Stringr manual[6]:

```
strings <- c(" 219 733 8965", "329-293-8753 ", "banana",
 "595 794 7569", "387 287 6718", "apple", "233.398.9187 ",
 "482 952 3315", "239 923 8115 and 842 566 4692",
 "Work: 579-499-7527", "$1000","Home: 543.355.3679")
```

```
phone <- "([2-9][0-9]{2})[- .]([0-9]{3})[- .]([0-9]{4})"
```

```
str_extract(strings, phone)
```

```
## [1] "219 733 8965" "329-293-8753" NA            "595 794 7569" "387
287 6718"
## [6] NA             "233.398.9187" "482 952 3315" "239 923 8115" "579-
499-7527"
## [11] NA             "543.355.3679"
```

```
bk.string.match <-  str_match(strings, phone)
bk.string.match
```

```
##        [,1]             [,2]   [,3]   [,4]
## [1,] "219 733 8965"   "219"  "733"  "8965"
## [2,] "329-293-8753"   "329"  "293"  "8753"
## [3,] NA               NA     NA     NA
## [4,] "595 794 7569"   "595"  "794"  "7569"
## [5,] "387 287 6718"   "387"  "287"  "6718"
## [6,] NA               NA     NA     NA
## [7,] "233.398.9187"   "233"  "398"  "9187"
## [8,] "482 952 3315"   "482"  "952"  "3315"
## [9,] "239 923 8115"   "239"  "923"  "8115"
## [10,] "579-499-7527"  "579"  "499"  "7527"
## [11,] NA              NA     NA     NA
## [12,] "543.355.3679"  "543"  "355"  "3679"
```

---

[6]https://cran.r-project.org/web/packages/stringr/stringr.pdf. p. 16, accessed on February 16, 2021.

Fields are conveniently separated so that you could, for example, parse out area codes.

2.3.1 Extract/Match All

```
str_extract_all(strings, phone)
```

```
## [[1]]
## [1] "219 733 8965"
##
## [[2]]
## [1] "329-293-8753"
##
## [[3]]
## character(0)
##
## [[4]]
## [1] "595 794 7569"
##
## [[5]]
## [1] "387 287 6718"
##
## [[6]]
## character(0)
##
## [[7]]
## [1] "233.398.9187"
##
## [[8]]
## [1] "482 952 3315"
##
## [[9]]
## [1] "239 923 8115" "842 566 4692"
##
## [[10]]
## [1] "579-499-7527"
##
```

```
## [[11]]
## character(0)
##
## [[12]]
## [1] "543.355.3679"
```

Five patterns:

```
bk.x <- c("<a> <b>", "<a> <>", "<a>", "", NA)
```

Matched against one pattern:

```
bk.x.wildcard <- str_match(bk.x, "<(.*?)> <(.*?)>")
bk.x.wildcard
```

```
##        [,1]        [,2] [,3]
## [1,] "<a> <b>" "a"  "b"
## [2,] "<a> <>"  "a"  ""
## [3,] NA        NA   NA
## [4,] NA        NA   NA
## [5,] NA        NA   NA
```

Here's another example from doracheng@maxchip.com.tw:[7]

```
str_match("2PIN Yield: 0.9933,1,1,0.9933,1,1,1,1","2PIN Yield: (.+?),")[,1]
```

```
## [1] "2PIN Yield: 0.9933,"
```

```
str_match("2PIN Yield: 0.9933,1,1,0.9933,1,1,1,1","2PIN Yield: (.+?),")[,2]
```

```
## [1] "0.9933"
```

Note the period/question mark wildcard usage:

```
str_match_all(bk.x, "<(.*?)>")
```

```
## [[1]]
##      [,1] [,2]
## [1,] "<a>" "a"
```

---

[7]"Str_match Function | R Documentation," accessed on January 30, 2021, www.rdocumentation.
org/packages/stringr/versions/1.4.0/topics/str_match.

```
## [2,] "<b>" "b"
##
## [[2]]
##       [,1]  [,2]
## [1,] "<a>" "a"
## [2,] "<>"  ""
##
## [[3]]
##       [,1]  [,2]
## [1,] "<a>" "a"
##
## [[4]]
##       [,1] [,2]
##
## [[5]]
##       [,1] [,2]
## [1,] NA    NA
```

```
str_extract(bk.x, "<.*?>")
```

```
## [1] "<a>" "<a>" "<a>" NA    NA
```

```
str_extract_all(bk.x, "<.*?>")
```

```
## [[1]]
## [1] "<a>" "<b>"
##
## [[2]]
## [1] "<a>" "<>"
##
## [[3]]
## [1] "<a>"
##
## [[4]]
## character(0)
##
## [[5]]
## [1] NA
```

Regular expressions use the following conventions:

- ABC: # match literal characters 'ABC:'

- \([a-zA-Z]+\) # one or more letters inside of parentheses

- * # zero or more spaces

- (.+) # capture one or more of any character (except newlines)

2.3.2 Case Functions

```
bk.text <- "Athenian citizen"
bk.text.upper.case <- str_to_upper(bk.text)
bk.text.upper.case
```

```
## [1] "ATHENIAN CITIZEN"
```

```
bk.text.lower.case <- str_to_lower(bk.text)
bk.text.lower.case
```

```
## [1] "athenian citizen"
```

```
bk.text.title <- str_to_title(bk.text)
bk.text.title
```

```
## [1] "Athenian Citizen"
```

```
bk.default.english.locale <- str_to_upper(bk.text,
    locale = "en")
bk.default.english.locale
```

```
## [1] "ATHENIAN CITIZEN"
```

Now try the same function with a different locale:

```
bk.using.not.English.locale <- str_to_upper(bk.text,
    locale = "ky-KG")
bk.using.not.English.locale
```

```
## [1] "ATHENIAN CITIZEN"
```

There is no difference from "en" in this case.

2.3.3 Geographically Aware (Locale-Aware) Functions

coll() is useful for locale-aware case-insensitive matching. Note that "\u0130" is Unicode for a Latin capital I with a dot above it. Not your everyday, gum-chewing letter.

There are three different codes set up here:

```
bk.i <- c("I", "\u0130", "i")
length(bk.i)
```

```
## [1] 3
```

```
bk.i
```

```
## [1] "I" "I" "i"
```

Note there are three types of "I."
Is there an "I" or "i" and if so, where is it?

```
str_detect(bk.i, fixed("i", TRUE))
```

```
## [1]  TRUE FALSE  TRUE
```

```
str_detect(bk.i, coll("i", TRUE))
```

```
## [1]  TRUE FALSE  TRUE
```

Same result since "coll" assumes "en" as locale:

```
str_detect(bk.i, coll("i", TRUE, locale = "tr"))  #now test using locale =
"tr" (Turkish).
```

```
## [1] FALSE  TRUE  TRUE
```

In this case, the Turkey locale treats the Latin I as a valid i (TRUE for middle element).

2.3.4 Combine Multiple Strings

Use the str_c function.

```
bk.letters <- str_c("Letter: ", letters)
bk.letters.with.separator <- str_c("Letter", letters,
   sep = ": ")
bk.letters.with.separator
```

```
##  [1] "Letter: a" "Letter: b" "Letter: c" "Letter: d" "Letter: e" "Letter: f"
##  [7] "Letter: g" "Letter: h" "Letter: i" "Letter: j" "Letter: k" "Letter: l"
## [13] "Letter: m" "Letter: n" "Letter: o" "Letter: p" "Letter: q" "Letter: r"
## [19] "Letter: s" "Letter: t" "Letter: u" "Letter: v" "Letter: w" "Letter: x"
## [25] "Letter: y" "Letter: z"
```

```
bk.combine.characters <- str_c(letters, " is for", "...")
bk.combine.characters
```

```
##  [1] "a is for..." "b is for..." "c is for..." "d is for..." "e is for..."
##  [6] "f is for..." "g is for..." "h is for..." "i is for..." "j is for..."
## [11] "k is for..." "l is for..." "m is for..." "n is for..." "o is for..."
## [16] "p is for..." "q is for..." "r is for..." "s is for..." "t is for..."
## [21] "u is for..." "v is for..." "w is for..." "x is for..." "y is for..."
## [26] "z is for..."
```

```
str_c(letters, " is for", "...")
```

```
##  [1] "a is for..." "b is for..." "c is for..." "d is for..." "e is for..."
##  [6] "f is for..." "g is for..." "h is for..." "i is for..." "j is for..."
## [11] "k is for..." "l is for..." "m is for..." "n is for..." "o is for..."
## [16] "p is for..." "q is for..." "r is for..." "s is for..." "t is for..."
## [21] "u is for..." "v is for..." "w is for..." "x is for..." "y is for..."
## [26] "z is for..."
```

```
bk.sequence <- str_c(letters[-26], " comes before ",
  letters[-1])
```

Everything except z "comes before" everything except a:

```
bk.sequence
```

```
##  [1] "a comes before b" "b comes before c" "c comes before d" "d comes
##      before e"
##  [5] "e comes before f" "f comes before g" "g comes before h" "h comes
##      before i"
##  [9] "i comes before j" "j comes before k" "k comes before l" "l comes
##      before m"
```

```
## [13] "m comes before n" "n comes before o" "o comes before p" "p comes
      before q"
## [17] "q comes before r" "r comes before s" "s comes before t" "t comes
      before u"
## [21] "u comes before v" "v comes before w" "w comes before x" "x comes
      before y"
## [25] "y comes before z"
```

```
bk.collapsed.letters <- str_c(letters, collapse = "")
bk.collapsed.letters
```

No spaces between letters, separated by null:

```
## [1] "abcdefghijklmnopqrstuvwxyz"
```

```
bk.comma.collapsed <- str_c(letters, collapse = ", ")
bk.comma.collapsed
```

```
## [1] "a, b, c, d, e, f, g, h, i, j, k, l, m, n, o, p, q, r, s, t, u, v,
w, x, y, z"
```

Missing inputs give missing outputs:

```
bk.missing <- str_c(c("a", NA, "b"), "-d")
bk.missing
```

```
## [1] "a-d" NA      "b-d"
```

Use str_replace_NA to display literal NAs:

```
bk.replace.na <- str_c(str_replace_na(c("a", NA, "b")), "-d")
bk.replace.na
```

```
## [1] "a-d"  "NA-d" "b-d"
```

Change encoding of a string by using str_conv.
Here's an example from encoding stringi::stringi

```
bk.x <- rawToChar(as.raw(177))
bk.x #note: on a Mac, you may get a different code "\xb1"
```

```
## [1] "±"
```

...with the plus/minus symbol:

Polish "a with ogonek"

```
bk.polish <- str_conv(bk.x, "ISO-8859-2")
bk.polish  #on a Mac you will get a "q"
```

```
## [1] "a"
```

...appears like an a with a cedilla:

```
bk.plus.minus <- str_conv(bk.x, "ISO-8859-1") # Plus-minus
bk.plus.minus
```

```
## [1] "±"
```

In general, you will get different results with different ISO standards.

2.4 More Complex Matching

2.4.1 What Does Not Match (Invert-Match)

This next function reminds me of effective advertisers. They show you things you didn't even know you need. But when you need them, it comes in handy.

```
bk.numbers <- "1 and 2 and 4 and 456"
bk.num.loc <- str_locate(bk.numbers, "[0-9]+")[[1]]
bk.num.loc   #output - integer
```

```
## [1] 1
```

```
class(bk.num.loc)
```

```
## [1] "integer"
```

```
bk.numbers <- "1 and 2 and 4 and 456"
bk.num.loc <- str_locate_all(bk.numbers, "[0-9]+")[[1]]
bk.num.loc
```

```
##      start end
## [1,]     1   1
## [2,]     7   7
```

```
## [3,]    13   13
## [4,]    19   21
```

```
bk.numbers.as.they.occur.in.the.string <- str_sub(bk.numbers,
      bk.num.loc[, "start"], bk.num.loc[, "end"])
bk.numbers.as.they.occur.in.the.string
```

```
## [1] "1"    "2"    "4"    "456"
```

```
bk.text.but.no.numbers <- invert_match(bk.num.loc)
```

```
bk.show.text <- str_sub(bk.numbers,
   bk.text.but.no.numbers[, "start"],
   bk.text.but.no.numbers[, "end"])
```

For example, if I had someone going to the grocery store for me, I might say "stock up on groceries using the invert-match of beets." Beets are the nauseous cousin of decent vegetables, fit only for horses and politicians.

```
bk.show.text
```

```
## [1] ""        " and " " and " " and " ""
```

^ matches the start of the string; $ matches the end of the string.

An easy mnemonic (https://rals.had.co.nzistrings.html): According to Dr. Hadley Wickham, "To remember which is which, try this mnemonic which I learned from Evan Misshula: if you begin with power (^), you end up with money ($)."

```
if (!require("htmlwidgets")) install.packages("htmlwidgets")
bk.view1 <- str_view(c("abc", "def", "fgh"), "[aeiou]")
bk.view1
```

See Figure 2-1 for output from bk.view1.

```
class(bk.view1)
```

```
## [1] "str_view"    "htmlwidget"
```

```
print(bk.view1)
```

Figure 2-1 shows the results of the print function. It appears in the selected output characters are highlighted. a and e are the only vowels.

Figure 2-1. *Print function displays in the lower right-hand screen, under "Viewer."*

Show start of a string:

```
bk.view2 <- str_view(c("abc", "def", "fgh"), "^")
print(bk.view2)
# |abc
# |def
# |fgh
```

Show first two characters:

```
bk.view3 <- str_view(c("abc", "def", "fgh"), "..")
print(bk.view3)
# ab
# de
# fg
```

Show all matches with str_view_all:

```
bk.view4 <- str_view_all(c("abc", "def", "fgh"), "d|e")
print(bk.view4)

# de
```

Use match to control what is shown:

```
str_view(c("abc", "def", "fgh"), "d|e")

# d      -- start of def
```

```
str_view(c("abc", "def", "fgh"), "d|e", match = TRUE)
# d
```

Note in the following that nothing is highlighted:

```
str_view(c("abc", "def", "fgh"), "d|e", match = FALSE)
```

Now highlighted characters are shown:

```
str_view(c("hair", "haematic"), "ha(i|e)")
# hai
# hae
```

2.5 Convenient Word Wrapping

Wrap strings into nicely formatted paragraphs. This function can save a lot of time if you need to clean up text which has been imported or pasted from many sources with a variety of formats. Due to space limitations, some of the following lists have been truncated[8]:

```
thanks_path <- file.path(R.home("doc"), "THANKS")
thanks <- str_c(readLines(thanks_path), collapse = "\n")
thanks <- word(thanks, 1, 3, fixed("\n\n"))
cat(str_wrap(thanks), "\n")

## R would not be what it is today without the invaluable help of these
people
## outside of the (former and current) R Core team, who contributed by
donating
## code, bug fixes and documentation: Valerio Aimale, Suharto Anggono,
Thomas
## Baier, Gabe Becker, Henrik Bengtsson, Roger Bivand, Ben Bolker, David
Brahm,
```

[8]https://cran.r-project.org/web/packages/stringr/stringr.pdf, p. 29, accessed on February 16, 2021.

----- truncated ----

List truncated to save space:

```
cat(str_wrap(thanks, width = 40), "\n")
```

```
## R would not be what it is today without
## the invaluable help of these people
## outside of the (former and current) R
## Core team, who contributed by donating
## code, bug fixes and documentation:
```

----- truncated ----

```
cat(str_wrap(thanks, width = 60, indent = 2), "\n")
```

```
## R would not be what it is today without the invaluable
## help of these people outside of the (former and current)
## R Core team, who contributed by donating code, bug fixes
## and documentation: Valerio Aimale, Suharto Anggono, Thomas
## Baier, Gabe Becker, Henrik Bengtsson, Roger Bivand, Ben
## Bolker, David Brahm, G"oran Brostr"om, Patrick Burns,
## Vince Carey, Saikat DebRoy, Matt Dowle, Brian D'Urso,
## Lyndon Drake, Dirk Eddelbuettel, Claus Ekstrom, Sebastian
```

----- truncated ----

```
cat(str_wrap(thanks, width = 60, exdent = 2), "\n")
```

```
## R would not be what it is today without the invaluable help
##    of these people outside of the (former and current) R
##    Core team, who contributed by donating code, bug fixes
##    and documentation: Valerio Aimale, Suharto Anggono, Thomas
##    Baier, Gabe Becker, Henrik Bengtsson, Roger Bivand, Ben
##    Bolker, David Brahm, G"oran Brostr"om, Patrick Burns,
##    Vince Carey, Saikat DebRoy, Matt Dowle, Brian D'Urso,
```

----- truncated ----

```
cat(str_wrap(thanks, width = 0, exdent = 2), "\n")
```

```
## R
##    would
##    not
##    be
##    what
##    it
##    is
##    today
```

----- truncated ----

```
cat(str_wrap(thanks, width = 40), "\n")
```

```
## R would not be what it is today without
## the invaluable help of these people
## outside of the (former and current) R
## Core team, who contributed by donating
## code, bug fixes and documentation:
```

----- truncated ----

```
cat(str_wrap(thanks, width = 60, indent = 2), "\n")
```

```
##    R would not be what it is today without the invaluable
## help of these people outside of the (former and current)
## R Core team, who contributed by donating code, bug fixes
## and documentation: Valerio Aimale, Suharto Anggono, Thomas
## Baier, Gabe Becker, Henrik Bengtsson, Roger Bivand, Ben
```

----- truncated ----

```
cat(str_wrap(thanks, width = 0, exdent = 2), "\n")
```

```
## R
##    would
##    not
##    be
```

```
##   what
##   it
##   is
```

----- truncated ----

2.6 Cleanup and Padding

2.6.1 Pad Your String, Not Your Expense Account

Padding, using str_pad, works like you would expect. Spaces or other characters are added on as specified:

```
bk.row.bind <-  rbind(
  str_pad("hadley", 30, "left"),
  str_pad("hadley", 30, "right"),
  str_pad("hadley", 30, "both"))
bk.row.bind
```

```
##        [,1]
## [1,] "                       hadley"
## [2,] "hadley                       "
## [3,] "              hadley           "
```

```
str_pad(c("a", "abc", "abcdef"), 10)
```

```
## [1] "         a" "       abc" "    abcdef"
```

```
str_pad("a", c(5, 10, 20))
```

```
## [1] "    a"          "         a"          "                   a"
```

```
str_pad("a", 10, pad = c("-", "_", " "))
```

```
## [1] "---------a" "_____a" "         a"
```

Longer strings are returned unchanged:

```
str_pad("hadley", 3)
```

```
## [1] "hadley"
```

2.6.2 Trim Whitespace from a String Using str_trim

See match characters in the book (e.g., \t and \n):

```
str_trim(" String with trailing and leading white space\t")
## [1] "String with trailing and leading white space"
```

Match characters not shown in the output:

```
str_trim("\n\nString with trailing and leading white space\n\n")
## [1] "String with trailing and leading white space"
str_squish(" String with trailing, middle, and leading white space\t")
## [1] "String with trailing, middle, and leading white space"
```

2.6.3 Remove All Whitespace from a String

```
str_squish("\n\nString with excess, trailing and leading white    space\n\n")
## [1] "String with excess, trailing and leading white space"
```

2.6.4 Truncate a String

Truncate a character string using str_truc.

```
bk.x <- "The stock market can remain irrational longer than you can remain
solvent. Keynes"
bk.right.left.center <- rbind(
  str_trunc(bk.x, 10, "right"),
  str_trunc(bk.x, 10, "left"),
  str_trunc(bk.x, 10, "center"))
bk.right.left.center

##      [,1]
## [1,] "The sto..."
## [2,] "... Keynes"
## [3,] "The ...nes"
```

```
nchar(bk.right.left.center)
```

```
##        [,1]
## [1,]    10
## [2,]    10
## [3,]    10
```

2.7 Regular Expressions with Stringr

Here's what Wikipedia says about regular expressions:

> *A regular expression, RegEx or regexp[i] (sometimes called a rational expression)[2][:)] is a sequence of characters that define a search pattern. Usually this pattern is used by string searching algorithms for "find" or "find and replace" operations on strings, or for input validation. It is a technique that developed in theoretical computer science and formal language theory.*
>
> *The concept arose in the 1950s when the American mathematician Stephen Cole Kleene formalized the description of a regular language. The concept came into common use with Unix text-processing utilities. Since the 1980s, different syntaxes for writing regular expressions exist, one being the POSIX standard and another, widely used, being the Perl syntax.*

Regular expressions are used in search engines, in search and replace dialogs of word processors and text editors, in text-processing utilities such as sed and AWK, and in lexical analysis. Many programming languages provide RegEx capabilities, built-in or via libraries. Since regular expressions are used in many programming languages, there are plenty of people who know the rather arcane structures thoroughly and find them handy. Stringr supports regular expressions. See this site for a more in-depth explanation of regular expressions:

`www.rdocumentation.org/packages/base/versions/3.5.3/topics/regex`

RStudio provides a cheat sheet for regular expressions in R.[9]

---

[9]`www.rstudio.com/wp-content/uploads/2016/09/RegExCheatsheet.pdf`. Accessed on January 30, 2021.

The following example is courtesy of Ian Kopacka and listed in the RStudio R regular expression cheat sheet:

```
bk.string <- c("Hiphopopotamus", "Rhymenoceros", "time for bottomless
lyrics")
bk.pattern <- "t.m"
bk.string.extract.first.match <- str_extract(bk.string,bk.pattern)
bk.string.extract.first.match
```

```
## [1] "tam" NA     "tim"
```

The pattern "t.m" means find a character string that starts with a "t" and ends with an "m". In this case, the "tam" in "Hiphopopotamus" is a match, and the "tim" in "time for bottomless lyrics" is a match. Nothing matches in the middle word, so the str_extract function outputs a NA for that term.

2.7.1 Regular Expression Variations

```
str_extract_all("The Cat in the Hat", "[a-z]+")
```

The "+" in the preceding expression indicates a repeat character.

```
## [[1]]
## [1] "he"  "at"  "in"  "the" "at"
```

```
str_extract_all("The Cat in the Hat", regex("[a-z]+", TRUE))
```

```
## [[1]]
## [1] "The" "Cat" "in"  "the" "Hat"
str_extract_all("a\nb\nc", "^.")
```

```
## [[1]]
## [1] "a"
```

```
str_extract_all("a\nb\nc", regex("^.", multiline = TRUE))
```

```
## [[1]]
## [1] "a" "b" "c"
```

Note   the line terminators /n are missing in the output.

```
str_extract_all("a\nb\nc", "a.")
```

```
## [[1]]
## character(0)
```

dotall says match everything and include special meta-characters such as \n (line terminator):

```
str_extract_all("a\nb\nc", regex("a.", dotall = TRUE))
```

```
## [[1]]
## [1] "a\n"
```

We asked for "a something" and got "a\n" meaning "a line terminator."

The following are Harvard sentences, from Darius Kazemi.[10] They are used as practice length sentences:

```
sentences[1:5]
```

```
##[1] "The birch canoe slid on the smooth planks."  "Glue the sheet to the
dark blue background."
##[3] "It's easy to tell the depth of a well."     "These days a chicken
leg is a rare dish."
##[5] "Rice is often served in round bowls."
```

```
fruit[1: 5]
```

```
##[1] "apple"      "apricot"    "avocado"     "banana"     "bell pepper"
```

```
length(words)
```

```
## [1] 980
```

```
words[1:5]
```

---

[10]"Stringr-Data: Sample Character Vectors for Practicing String Manipulations. In Stringr: Simple, Consistent Wrappers for Common String Operations," accessed on January 30, 2021, https://rdrr.io/cran/stringr/man/stringr-data.html.

```
## [1] "a"          "able"      "about"     "absolute" "accept"
```

```
head(sentences,10)
```

```
##[1] "The birch canoe slid on the smooth planks."  "Glue the sheet to the
dark blue background."
##[3] "It's easy to tell the depth of a well."      "These days a chicken
leg is a rare dish."
##[5] "Rice is often served in round bowls."        "The juice of lemons
makes fine punch."
##[7] "The box was thrown beside the parked truck." "The hogs were fed
chopped corn and garbage."
##[9] "Four hours of steady work faced us."         "Large size in
stockings is hard to sell."
```

Stringr is a powerful package which enables you to work with messy character data and extract or identify what you need. When you reach a certain level of complexity, regular expressions (inside Stringr) function as condensed code to get just what you need. They sometimes look a bit like hieroglyphics, but once nailed down, you can use them the rest of your career (in many languages, not just R). A later chapter in this book will cover the basics of regular expressions (often called "RegEx").

CHAPTER 3

Lubridate: Date and Time Processing

Lubridate starts out answering a simple question—is it AM or PM, based on a date and hour? From there, it gets more complex but maintains a consistent approach to working with dates, times, and the combination date-times.

3.1 Hard-Coding Coffee Time

```
if (!require("lubridate")) install.packages("lubridate")

bk.coffee.time <- as.POSIXct("080418 10:11",
  format = "%y%m%d %H:%M")
am(bk.coffee.time)

## [1] TRUE

bk.time.span <- interval(ymd("2019-01-01"), ymd("2019-06-30"))
bk.time.span.value <- as.duration(bk.time.span)
bk.time.span.value

## [1] "15552000s (~25.71 weeks)"
```

© William Yarberry 2021
W. Yarberry, *CRAN Recipes*, https://doi.org/10.1007/978-1-4842-6876-6_3

3.2 Duration Calculations

Note the following output is in seconds and in character format:

```
bk.how.long.in.hours.and.minutes <- duration(hours = 4,
    minutes = 10)
bk.how.long.in.hours.and.minutes
## [1] "15000s (~4.17 hours)"
```

```
bk.hours.numeric <-
    as.numeric(bk.how.long.in.hours.and.minutes,"hours")
bk.hours.numeric
```

```
## [1] 4.166667
```

```
bk.minutes.numeric <-
    as.numeric(bk.how.long.in.hours.and.minutes,"minutes")
bk.minutes.numeric
```

```
## [1] 250
```

Show the format of five days:

```
bk.five.whole.days <- as.period(5, unit = "day") #five days
bk.five.whole.days
```

```
## [1] "5d 0H 0M 0S"
```

3.3 Spanning Two Dates Using Interval

How to span two dates—2009 vs. 2010? This syntax shows one of each time measure—year, month, day, hour, minute, and second:

```
bk.span.of.two.dates <- interval(ymd_hms("2009-01-01 00:00:00"),
    ymd_hms("2010-02-02 01:01:01"))
as.period(bk.span.of.two.dates)
```

```
## [1] "1y 1m 1d 1H 1M 1S"
```

CHAPTER 3   LUBRIDATE: DATE AND TIME PROCESSING

Change the span to be between two years, same day:

```
bk.span.of.two.dates <- interval(ymd_hms("2009-01-01 00:00:00"),
  ymd_hms("2010-01-01 00:00:00"))
as.period(bk.span.of.two.dates, unit = "day")
```

```
## [1] "365d 0H 0M 0S"
```

Show the span between a leap year and a non-leap year:

```
bk.leap.year.span <- interval(ymd_hms("2016-01-01 00:00:00"),
  ymd_hms("2017-01-01 01:00:00"))
as.period(bk.leap.year.span, unit = "day")
```

```
## [1] "366d 1H 0M 0S"
```

3.3.1 Work with Time Zones

```
bk.chicago.time.interval <- interval(ymd("2016/11/06",
 tz = "America/Chicago"),ymd("2016/11/09"))
#you can also enter date as 2016-11-06
bk.hours <- as.period(bk.chicago.time.interval, unit = "hours")
bk.hours
```

```
## [1] "67H 0M 0S"
```

Now use Belfast, Ireland, as time zone:

```
bk.europe.belfast.time.interval <- interval(ymd("2016/11/06",
  tz = "Europe/Belfast"),ymd("2016/11/09"))
bk.hours <- as.period(bk.europe.belfast.time.interval,
  unit = "hours")
bk.hours
```

```
## [1] "72H 0M 0S"
```

Duration: how much time between two points in time? In this syntax, the %--% operator creates a date interval by subtracting start from end dates. Note that the calculated variable, bk.interval, includes two date/times, start and finish.

```
bk.start.date.time <- mdy_hm("8-11-2017 5:21",
  tz = "US/Eastern")
bk.end.date.time  <- mdy_hm("8-12-2018 6:21",
  tz = "US/Eastern")
bk.interval <- bk.start.date.time %--% bk.end.date.time
bk.interval

## [1] 2017-08-11 05:21:00 EDT--2018-08-12 06:21:00 EDT
```

Note that bk.interval is special class:

```
class(bk.interval)
[1] "Interval"
attr(,"package")
[1] "lubridate"
```

3.4 Calculate Duration (Seconds) Between Two Date/Times

Calculate the duration from the properly formatted interval:

```
bk.duration <- as.duration(bk.interval)
bk.duration

## [1] "31626000s (~1 years)"
```

Here's another example from history. Frederick III upgraded Prussia from a duchy to a kingdom on January 18, 1701. The Allied Control Council, on February 25, 1947, after the Second World War, declared that the state of Prussia "hereby ceases to exist." How many years did Prussia, as a recognized nation, exist?

```
bk.prussia.started <- ymd("1701-01-18")
bk.prussia.ended <- ymd("1947-02-25")
bk.interval <- bk.prussia.started %--% bk.prussia.ended
```

```
bk.prussia.in.existence <- as.duration(bk.interval)
bk.prussia.in.existence
```

```
## [1] "7766236800s (~246.1 years)"
```

Prussia lasted a little more than 246 years.

When using the function as.duration, the output is in seconds.

```
bk.seventeen.seconds <- as.duration(17)
bk.seventeen.seconds
```

```
## [1] "17s"
```

To show results with a different time metric, use hours or minutes as a parameter:

```
as.numeric(bk.seventeen.seconds, "hours")
```

```
## [1] 0.004722222
as.numeric(bk.seventeen.seconds, "minutes")
```

```
## [1] 0.2833333
```

Note that "duration" and "as.duration" are different functions:

```
bk.hours <- 25
bk.minutes <- 8
bk.twenty.five.hours <- duration(hours = bk.hours)
bk.twenty.five.hours
```

```
## [1] "90000s (~1.04 days)"
```

Specify hours and minutes as duration parameters:

```
bk.twenty.five.hours.and.eight.minutes <-
    duration(hours = bk.hours, minutes = bk.minutes)
bk.twenty.five.hours.and.eight.minutes
```

```
## [1] "90480s (~1.05 days)"
```

Show the interval as.duration:

```
bk.time.span <- interval(ymd("2019-01-01"), ymd("2019-06-30")) bk.time.
span.value <- as.duration(bk.time.span)
bk.time.span.value
```

```
##[1] "15552000s (~25.71 weeks)"
```

Show duration in hours:

```
dur <- duration(hours = 10, minutes = 6)
as.numeric(dur, "hours")
```

```
## [1] 10.1
```

Show duration in minutes:

```
as.numeric(dur, "minutes")
```

```
## [1] 606
```

3.5 More Interval Calculations

The interval function shows the end date first:

```
interval(ymd(20090201), ymd(20090101))
```

```
## [1] 2009-02-01 UTC--2009-01-01 UTC
```

```
bk.date1 <- ymd_hms("2009-03-08 01:59:59")
bk.date2 <- ymd_hms("2000-02-29 12:00:00")
bk.interval <- interval(bk.date2, bk.date1)
bk.interval
```

```
## [1] 2000-02-29 12:00:00 UTC--2009-03-08 01:59:59 UTC
```

This is an alternate way to calculate the interval:

```
bk.interval.alternate.method <-
  lubridate::`%--%`(lubridate::ymd("2009-03-08"),
  lubridate::ymd("2000-02-29"))
bk.interval.alternate.method
```

```
## [1] 2009-03-08 UTC--2000-02-29 UTC
```

Switch the beginning and end dates:

```
bk.interval.negative <-interval(bk.date1, bk.date2)
bk.interval.negative
```

```
## [1] 2009-03-08 01:59:59 UTC--2000-02-29 12:00:00 UTC
```

Show a single month's span:

```
bk.span <- interval(ymd(20090101), ymd(20090201))
bk.span
```

```
## [1] 2009-01-01 UTC--2009-02-01 UTC
```

Show an interval in seconds:

```
bk.interval.length.in.seconds <- int_length(bk.span)
bk.interval.length.in.seconds
```

```
## [1] 2678400
```

As a sanity check, compare the interval in seconds calculated earlier with a direct calculation of 31 days * 24 hours * 60 minutes * 60 seconds:

```
bk.comparison <- 31*24*60*60
bk.comparison
```

```
## [1] 2678400
```

Interval start time:

```
bk.start <- int_start(bk.span)
bk.start
```

```
## [1] "2009-01-01 UTC"
```

Interval stop time:

```
bk.stop <- int_end(bk.span)
bk.stop
```

```
## [1] "2009-02-01 UTC"
```

What does bk.span look like when start and end are reversed? It shows the mirror image:

```
bk.span.flipped <- int_flip(bk.span)
bk.span.flipped
```

```
## [1] 2009-02-01 UTC--2009-01-01 UTC
```

3.6 Interval Overlaps

Create a new interval.

End date:

```
bk.date1.second.range <- ymd_hms("2010-06-08 01:59:59")
```

Beginning date:

```
bk.date2.second.range <- ymd_hms("2010-04-29 12:00:00")
bk.another.span <- interval(bk.date1.second.range,
  bk.date2.second.range)
bk.another.span
```

```
## [1] 2010-06-08 01:59:59 UTC--2010-04-29 12:00:00 UTC
```

Does this second span overlap with the first span? No, because the second interval is set in 2010:

```
bk.overlap.or.not <- int_overlaps(bk.another.span,bk.span)
bk.overlap.or.not
```

```
## [1] FALSE
```

Now change the range so that they do overlap.

End date:

```
bk.date1.second.range <- ymd_hms("2009-06-08 01:59:59")
```

Beginning date:

```
bk.date2.second.range <- ymd_hms("2008-04-29 12:00:00")
bk.another.span <- interval(bk.date1.second.range,
```

```
  bk.date2.second.range)
bk.another.span
```

```
## [1] 2009-06-08 01:59:59 UTC--2008-04-29 12:00:00 UTC
```

Does this second span overlap with the first span?

```
bk.overlap.or.not <- int_overlaps(bk.another.span,bk.span)
bk.overlap.or.not
```

```
## [1] TRUE
```

3.7 Interval Shift

Calculate an interval shift. Use this when the entire schedule moves forward or backward a certain amount of time[1]:

```
int <- interval(ymd("2001-01-01"), ymd("2002-01-01"))
int_shift(int, duration(days = 11))
```

```
## [1] 2001-01-12 UTC--2002-01-12 UTC
```

```
int_shift(int, duration(hours = -1))
```

```
## [1] 2000-12-31 23:00:00 UTC--2001-12-31 23:00:00 UTC
```

Flip dates on their head:

```
bk.interval <- interval(ymd("2001-01-01"), ymd("2002-11-11"))
bk.interval
```

```
## [1] 2001-01-01 UTC--2002-11-11 UTC
```

```
bk.interval.flip <- int_flip(bk.interval)
bk.interval.flip
```

```
## [1] 2002-11-11 UTC--2001-01-01 UTC
```

---

[1]www.rdocumentation.org/packages/lubridate/versions/1.5.6/topics/int_shift. Accessed on January 22, 2021.

3.8 Alignment

Dates and time measurement began in ancient history, beginning with the Sumerian sexagesimal system. Modifications have been required ever since. Even now, tiny adjustments must occasionally be made to correct for an imperfect rotation of the earth. Recent improvements in measurement technology enable changes in the earth's rotation rate due to earthquakes such as the ones in Sumatra.[2]

Do beginning or ending dates match? Lubridate provides a true/false, logical answer.

```
bk.date1 <- ymd_hms("2009-06-08 01:59:59")   #end date
bk.date2 <- ymd_hms("2008-04-29 12:00:00")   #beginning date
bk.span1 <- interval(bk.date1,bk.date2)
bk.date3 <- ymd_hms("2011-06-08 01:59:59")   #end date
bk.date4 <- ymd_hms("2008-04-29 12:00:00")   #beginning date
bk.span2 <- interval(bk.date1,bk.date2)
bk.aligns.true.or.false <- int_aligns(bk.span1, bk.span2)
bk.aligns.true.or.false
```

```
## [1] TRUE
```

The answer is true because both spans have a common beginning date.
Intervals calculated int-diff(dates):

```
bk.multiple.dates <- bk.date4 + days(1:5)
bk.multiple.dates
```

```
## [1] "2008-04-30 12:00:00 UTC" "2008-05-01 12:00:00 UTC"
## [3] "2008-05-02 12:00:00 UTC" "2008-05-03 12:00:00 UTC"
## [5] "2008-05-04 12:00:00 UTC"
```

Create a vector of five dates—April 30, May 1, May 2, May 3, and May 4:

```
bk.int.diff <- int_diff(bk.multiple.dates)
bk.int.diff
```

---

[2]Private correspondence with Michael Torbett, retired professor of astronomy at Macon State University. January 30, 2021.

```
## [1] 2008-04-30 12:00:00 UTC--2008-05-01 12:00:00 UTC
## [2] 2008-05-01 12:00:00 UTC--2008-05-02 12:00:00 UTC
## [3] 2008-05-02 12:00:00 UTC--2008-05-03 12:00:00 UTC
## [4] 2008-05-03 12:00:00 UTC--2008-05-04 12:00:00 UTC
```

3.9 Periods

Periods record changes in wall-clock time, for example, between when the TV show *The Big Bang Theory* starts and ends tonight. The length of a time unit, such as a year, will vary, depending on whether it is a standard year or a leap year. There might be other reasons as well. Seconds are constant, but any higher level of time, such as a minute, will vary depending on where it occurs in calendar time. If you work with precision timing on projects, understanding the nuances of R date-time calculations becomes critically important.

Create some periods that could be added to a date-time, so that the result is a valid start/stop date-time:

```
bk.ninety.seconds.and.five.minutes <- period(c(90, 5),
    c("second", "minute"))
bk.ninety.seconds.and.five.minutes
```

```
## [1] "5M 90S"
```

```
class(bk.ninety.seconds.and.five.minutes)
```

```
## [1] "Period"
## attr(,"package")
## [1] "lubridate"
```

```
bk.ninety.seconds.and.five.minutes
```

```
## [1] "5M 90S"
```

Some basic calculations:

```
period(-1, "days")
```

```
## [1] "-1d 0H 0M 0S"
```

```
period(c(3, 1, 2, 13, 1),
  c("second", "minute", "hour", "day", "week"))
```

```
## [1] "20d 2H 1M 3S"
```

```
period(c(1, -60), c("hour", "minute"))
```

```
## [1] "1H -60M 0S"
```

```
period(0, "second")
```

```
## [1] "0S"
```

3.10 Sequencing

Sometimes you want to calculate the next N periods. This includes minutes, hours, days, months, years, and so on.

```
bk.sequence.start <- ymd(190102)    #January 2, 2019
bk.next.12.months <- bk.sequence.start + months(1:12)
bk.next.12.months
```

```
##  [1] "2019-02-02" "2019-03-02" "2019-04-02" "2019-05-02" "2019-06-02"
##  [6] "2019-07-02" "2019-08-02" "2019-09-02" "2019-10-02" "2019-11-02"
## [11] "2019-12-02" "2020-01-02"
```

```
length(bk.next.12.months)
```

```
## [1] 12
```

Generate date sequences between specific days:

```
start1 <- mdy("06/01/20")
end1 <- mdy("06/10/20")
end2 <- mdy("09/30/25")
x <- seq(start1, end1,"days")
x
```

```
##  [1] "2020-06-01" "2020-06-02" "2020-06-03" "2020-06-04" "2020-06-05"
##  [6] "2020-06-06" "2020-06-07" "2020-06-08" "2020-06-09" "2020-06-10"
```

Generate a sequence of 24 months, by month:

```
y <- seq(start1, by = "month", length.out = 24)
y
```

```
##  [1] "2020-06-01" "2020-07-01" "2020-08-01" "2020-09-01" "2020-10-01"
##  [6] "2020-11-01" "2020-12-01" "2021-01-01" "2021-02-01" "2021-03-01"
## [11] "2021-04-01" "2021-05-01" "2021-06-01" "2021-07-01" "2021-08-01"
## [16] "2021-09-01" "2021-10-01" "2021-11-01" "2021-12-01" "2022-01-01"
## [21] "2022-02-01" "2022-03-01" "2022-04-01" "2022-05-01"
```

Use quarters to generate a sequence:

```
q <- seq(start1, end2,by = "quarter")
q
```

```
##  [1] "2020-06-01" "2020-09-01" "2020-12-01" "2021-03-01" "2021-06-01"
##  [6] "2021-09-01" "2021-12-01" "2022-03-01" "2022-06-01" "2022-09-01"
## [11] "2022-12-01" "2023-03-01" "2023-06-01" "2023-09-01" "2023-12-01"
## [16] "2024-03-01" "2024-06-01" "2024-09-01" "2024-12-01" "2025-03-01"
## [21] "2025-06-01" "2025-09-01"
```

Lubridate style parsing:

```
period("2M 1sec")
```

```
## [1] "2M 1S"
```

```
period("2hours 2minutes 1second")
```

```
## [1] "2H 2M 1S"
```

```
period("2d 2H 2M 2S")
```

```
## [1] "2d 2H 2M 2S"
```

```
period("2days 2hours 2mins 2secs")
```

```
## [1] "2d 2H 2M 2S"
```

Missing numerals default to 1. Repeated units are added up:

```
duration("day day")
```

```
## [1] "172800s (~2 days)"
```

3.11 Distinction Between Period and Duration

According to a University of Virginia website:[3]

- *An Interval is elapsed time in seconds between two specific dates. (If no time is provided, the time for each date is assumed to be 00:00:00, or midnight.)*

- *A Duration is elapsed time in seconds independent of a start date.*

- *A Period is elapsed time in "calendar" or "clock" time (4 weeks, 2 months, etc) independent of a start date.*

Keep these distinctions in mind when reviewing the following examples. "tz" is time zone:

```
bk.boundary <- ymd_hms("2009-03-08 01:59:59",
  tz="America/Chicago")
bk.boundary
```

```
## [1] "2009-03-08 01:59:59 CST"
```

```
bk.boundary2 <- bk.boundary + days(1) # period
bk.boundary2
```

```
## [1] "2009-03-09 01:59:59 CDT"
```

```
bk.boundary3 <- bk.boundary + ddays(1) # duration
bk.boundary3
```

```
## [1] "2009-03-09 02:59:59 CDT"
```

---

[3]"Working with Dates and Time in R Using the Lubridate Package | University of Virginia Library Research Data Services + Sciences," accessed on January 22, 2021, https://data.library.virginia.edu/working-with-dates-and-time-in-r-using-the-lubridate-package/.

```
bk.true.false <- is.period(as.Date("2009-08-03"))
bk.true.false
```

```
## [1] FALSE
```

Convert seconds to a period and a period to seconds. Periods vary a bit, so to convert to the exact measure of seconds, assumptions are made. The assumption is that there are 365.25 days in a calendar year and $365.25/12 = 30.4375$ days in a month.

Two days' worth of seconds = 48 hours x 60 minutes x 60 seconds = 172,800. The function seconds_to_period() converts a quantity of seconds back to a more human-readable count of days, hours, and so on.

```
bk.period.from.seconds <- seconds_to_period(172800)
bk.period.from.seconds
```

```
## [1] "2d 0H 0M 0S"
```

Set variable to 1 year and 1 month = 13 months:

```
bk.period.to.seconds <- period_to_seconds(years(1) + months(1))
bk.period.to.seconds
```

```
## [1] 34187400
```

```
bk.period.to.seconds.in.months <- bk.period.to.seconds /
    (60*60*24*30.4375)
bk.period.to.seconds.in.months
```

```
## [1] 13
```

3.12 Timespan

Timespan uses three measurements:

- Duration
- Interval
- Period

The Lubridate manual[4] defines these three measurements as follows:

- Durations record the exact number of seconds in a timespan. They measure the exact passage of time but do not always align with human measurements like hours, months, and years.

- Periods record the change in the clock time between two date-times. They are measured in human units: years, months, days, hours, minutes, and seconds.

- Intervals are timespans bound by two real date-times. Intervals can be accurately converted to periods and durations.

```
bk.duration <-duration(3690, "seconds")
bk.duration
```

```
## [1] "3690s (~1.02 hours)"
```

```
bk.period <- period(3690, "seconds")
bk.period
```

```
## [1] "3690S"
```

```
bk.define.second.minute.hour <- period(second = 30,
   minute = 1, hour = 1)
bk.define.second.minute.hour
```

```
## [1] "1H 1M 30S"
```

Start/stop actual calendar dates and wall-clock times:

```
bk.interval <- interval(ymd_hms("2009-08-09 13:01:30"),
   ymd_hms("2009-08-09 12:00:00"))
bk.interval
```

```
## [1] 2009-08-09 13:01:30 UTC--2009-08-09 12:00:00 UTC
```

---

[4]https://cran.r-project.org/web/packages/lubridate/lubridate.pdf, 57. Accessed on January 22, 2021.

This example is from the excellent website rdocumentation.org:[5]

```
date1 <- ymd_hms("2009-03-08 01:59:59")
date2 <- ymd_hms("2000-02-29 12:00:00")
interval(date2, date1)
```

```
## [1] 2000-02-29 12:00:00 UTC--2009-03-08 01:59:59 UTC
```

```
interval(date1, date2)
```

```
## [1] 2009-03-08 01:59:59 UTC--2000-02-29 12:00:00 UTC
```

Daylight savings time boundary. Adding days(1) puts the date/time exactly one day later:

```
bk.date <- ymd_hms("2009-03-08 01:59:59")
bk.one.more.day <- bk.date + days(1)
bk.date
```

```
## [1] "2009-03-08 01:59:59 UTC"
```

```
bk.one.more.day
```

```
## [1] "2009-03-09 01:59:59 UTC"bk.date + ddays(1)
```

Let's see if we can trick Lubridate by asking for exactly one calendar year from February 29:

```
bk.date2 <- ymd_hms("2000-02-29 12:00:00")
bk.one.year.later <- bk.date2 + years(1)
bk.one.year.later
```

```
## [1] NA
```

No, Lubridate knows about leap years and will not let you do this. From February 29, 2000, forward exactly one year would yield February 29, 2001, an invalid date. Hence, NA. When it comes to dates, Lubridate really is the smartest package in the room.

---

[5] "Interval Function | R Documentation," accessed on January 22, 2021, www.rdocumentation.org/packages/lubridate/versions/1.7.4/topics/interval.

Using dyears rather than years, we get a logical answer to the question of what date is one year after a leap year.

```
bk.date2 <- ymd_hms("2000-02-29 12:00:00")
bk.one.year.later <- bk.date2 + dyears(1)
bk.one.year.later
```

```
## [1] "2001-02-28 18:00:00 UTC"
```

Use this logic to show all months advanced by a month. NA is shown for months less than 31 days.

```
bk.date3 <- ymd_hms("2009-01-31 01:00:00")
bk.date3 + c(0:11) * months(1)
```

```
##  [1] "2009-01-31 01:00:00 UTC" NA
##  [3] "2009-03-31 01:00:00 UTC" NA
##  [5] "2009-05-31 01:00:00 UTC" NA
##  [7] "2009-07-31 01:00:00 UTC" "2009-08-31 01:00:00 UTC"
##  [9] NA                        "2009-10-31 01:00:00 UTC"
## [11] NA                        "2009-12-31 01:00:00 UTC"
```

Create an interval:

```
bk.span <- bk.date2 %--% bk.date
bk.span
```

```
## [1] 2000-02-29 12:00:00 UTC--2009-03-08 01:59:59 UTC
```

Advance by one year:

```
bk.date4 <- ymd_hms("2009-01-01 00:00:00")
bk.date5  <-bk.date4 + years(1)
bk.date5
```

```
## [1] "2010-01-01 UTC"
```

Go back three days and then advance six hours:

```
bk.date7  <- ymd_hms("2009-03-08 01:59:59") # DST boundary
bk.date8 <- bk.date7 - days(3) + hours(6)
bk.date8
```

```
## [1] "2009-03-05 07:59:59 UTC"
```

Crossing the **daylight savings time boundary**. Notice how Lubridate sends time backward here.

```
bk.date <- ymd_hms("2009-03-08 01:59:59")
bk.date9 <- bk.date + 3 * seconds(10)
bk.date   #start
```

```
## [1] "2009-03-08 01:59:59 UTC"
```

```
bk.date9 #finish
```

```
## [1] "2009-03-08 02:00:29 UTC"
```

Going forward by specified months and days. You can think of bk.date10 as a chunk of time:

```
bk.date10 <- months(6) + days(1)
bk.date10
```

```
## [1] "6m 1d 0H 0M 0S"
```

Now, add bk.date10 to the day I'm writing this, 2021-1-22:

```
date.in.future.6mo.1day <- ymd("2021-01-22") + bk.date10
date.in.future.6mo.1day
```

```
## [1] "2021-07-23"
```

Conversion of an interval stated as two calendar dates into a number of weeks. To verify results, multiply 52 weeks by 34.826 years to get 1811.286 weeks. Again, note that a week is not a precise number of time ticks, since during the timespan there were many leap years.[6]

```
bk.int <- interval(ymd("1980-01-01"), ymd("2014-09-18"))
bk.time.length.in.weeks <- time_length(bk.int, "week")
bk.time.length.in.weeks
```

```
## [1] 1811.286
```

---

[6]According to Wikipedia.org, "A week is defined as an interval of exactly seven days,[b] so that technically, except at daylight saving time transitions or leap seconds, 1 week = 7 days = 168 hours = 10,080 minutes = 604,800 seconds." https://en.wikipedia.org/wiki/Week, accessed on January 22, 2021.

Same calculation, using the year metric:

```
bk.year.age.at.last.whole.year <- time_length(bk.int, "year")
bk.year.age.at.last.whole.year
```

```
## [1] 34.71233
```

Use trunc (truncation) function to obtain whole years:

```
bk.truncated.time.length <- trunc(time_length(bk.int, "year"))
bk.truncated.time.length
```

```
## [1] 34
```

3.12.1 Contrast of Intervals and Durations

```
bk.int <- interval(ymd("1900-01-01"), ymd("1999-12-31"))
bk.int
```

```
## [1] 1900-01-01 UTC--1999-12-31 UTC
```

Show one year using time_length:

```
bk.time.interval.length <-  time_length(bk.int, "year")
bk.time.interval.length
```

```
## [1] 99.99726
```

```
bk.time.interval.as.year <- time_length(as.duration(bk.int),
  "year")
bk.time.interval.as.year
```

```
## [1] 99.99452
```

R's default starting date is January 1, 1970.

Show date as 44 days since January 1, 1970:[7]

```
bk.date <- as_date(44)
bk.date
```

```
## [1] "1970-02-14"
```

The Lubridate manual uses the word "**origin**" to mean starting date/time. I'm not sure why that word is used. The word origin comes from the French *origine*, from Latin *origo*, meaning "to rise." In this case, the etiology does not help much. At the risk of being considered a spoil-sport luddite, I prefer the simple Anglo-Saxon "beginning date and time" term.

Add 21 days to a starting date of February 1, 2019:

```
bk.date <- as_date(21, origin = lubridate::ymd("2019-02-01"))
bk.date
```

```
## [1] "2019-02-22"
```

Simple character to R date conversion:

```
bk.date <- as_datetime("2019-01-02")
bk.date
```

```
## [1] "2019-01-02 UTC"
```

```
class(bk.date)
```

```
## [1] "POSIXct" "POSIXt"
```

This example shows the importance of separators. Entering the year, month, and day without separators causes Lubridate to consider the entire "date" as a number of seconds. Those seconds are added to R's default starting date, January 1, 1970, giving the wrong result:

```
bk.date.only.number.input <- as_datetime(20190102)
bk.date.only.number.input
```

```
## [1] "1970-08-22 16:21:42 UTC"
```

---

[7]This reminds me of the Latin term "novus ordo seclorum," which is printed on the back of a US one dollar bill. It means "a new order of the ages." Apparently, the R designers considered their new order date as January 1, 1970.

3.12.2 **Coordinated Universal Time Zone (UTC)**

According to wikipedia.org, "Coordinated Universal Time or UTC is the primary time standard by which the world regulates clocks and time. It is within about 1 second of mean solar time at 0° longitude and is not adjusted for daylight saving time. It is effectively a successor to Greenwich Mean Time (GMT)."[8]

It is a good practice to immediately convert any dates read in from external sources into a valid R date. Using the R POSIXct and POSIXlt classes allow for dates and times with control for time zones. If you leave variables in character format, everything you do will need to be hand coded, with the added risk of getting bad results. Use R as R, not as a typewriter.

Unless otherwise stated, UTC is assumed by Lubridate:

```
bk.date_utc <- ymd_hms("2010-08-03 00:50:50")
```

```
## [1] "2010-08-03 00:50:50 UTC"
```

Now calculate a date using the Europe/London time zone:

```
bk.date_europe <- ymd_hms("2010-08-03 00:50:50",
    tz="Europe/London")
bk.date_europe
```

```
## [1] "2010-08-03 00:50:50 BST"
```

3.12.3 **as_date vs. as.Date**

To illustrate the difference, first show what the same date looks like using both functions. Without a time zone specified, there is no difference.

```
bk.as_date_versus_as.Date <- c(as_date(bk.date_utc),
    as.Date(bk.date_utc))
bk.as_date_versus_as.Date
```

```
## [1] "2010-08-03" "2010-08-03"
```

---

[8]"Coordinated Universal Time," in Wikipedia, January 19, 2021, https://en.wikipedia.org/w/index.php?title=Coordinated_Universal_Time&oldid=1001289270.

Now, using the same date, show that the value changes based on the specified time zone "Europe/London" as listed earlier:

```
bk.as_date_versus_as.Date <- c(as_date(bk.date_europe), as.Date(bk.date_
europe))
bk.as_date_versus_as.Date
```

```
## [1] "2010-08-03" "2010-08-02"
```

3.12.4 Create a Date/Time Object via Hard Coding

```
bk.date.and.time <- ymd_hms("2010-12-03 00:56:50")
bk.date.and.time
```

```
## [1] "2010-12-03 00:56:50 UTC"
```

```
class(bk.date.and.time)
```

```
## [1] "POSIXct" "POSIXt"
```

Extract date only, without time:

```
bk.date.only <- date(bk.date.and.time)
bk.date.only
```

```
## [1] "2010-12-03"
```

Get the date, based on a time zone. bk.data.and.time originally (without tz) had a date of December 3, 2010, but with the America/New_York tz, it has a value of December 2, 2010:

```
bk.date.only.with.time.zone <- as.Date(bk.date.and.time,
  tz = "America/New_York")
bk.date.only.with.time.zone
```

```
## [1] "2010-12-02"
```

3.12.5 Revise a Date by Individually Changing Month, Day, Year

```
bk.date <- ymd("2018-12-31")
bk.date.revised <- update(bk.date, month = 01,
  mday = 01, year = 2019)
bk.date.revised
```

```
## [1] "2019-01-01"
```

3.12.6 Fractional Year

Use June 30 as an estimate for a half year. Result of the calculation is 2017 plus a fraction close to 0.5:

```
bk.date <- ymd("2017-06-30")
bk.decimal <- decimal_date (bk.date)
bk.decimal
```

```
## [1] 2017.493
```

```
bk.show.difference <- bk.decimal - 2017
bk.show.difference
```

```
## [1] 0.4931507
```

If you subtract 2017 from the decimal date, you get approximately 1/2 year back again, with the calendar date of June 30, 2017. The date is in standard POSIX format:

```
bk.date.again <- date_decimal(bk.decimal)
bk.date.again
```

```
## [1] "2017-06-30 00:00:00 UTC"
```

```
class(bk.date.again)
```

```
## [1] "POSIXct" "POSIXt"
```

3.12.7 Work Day

In this case, day three is a Tuesday because the week starts on a Sunday:

```
bk.date <- as.Date("2019-01-15")
bk.working.day <- wday(bk.date)
bk.working.day
```

```
## [1] 3
```

3.12.8 Is Daylight Savings Time in Effect?

Is this date in daylight savings time for time zone "America/New_York"? The answer will be logical, true or false.

```
bk.date.time <- ymd("2019-03-10",tz = "America/New_York")
bk.daylight.savings.time <- dst(bk.date.time)
bk.daylight.savings.time
```

```
## [1] FALSE
```

3.12.9 Guess Formats

Guess formats routine for varieties of dates—how to handle a column of date-times when the formats vary from data element to data element. Guess formats is not an intuitive function. The idea is to match up character dates in a variety of formats, everything from "Thu, 1 July 2004 22:30:00" to "12/31/99". Then those dates get matched with possible formats, and a verbose output is provided. I hate to be a whiner, but at first glance, it just looks to stinkin' hard to use. What a relief when I saw simplifying code from stack overflow user agstudy[9]:

```
bk.dates.in.different.formats <- sampleDates <-
   c("4/6/2004","4/6/2004","4/6/2004","4/7/2004",
   "4/6/2004","4/7/2004","2014-06-28","2014-06-30","2014-07-12",
   "2014-07-29","2014-07-29","2014-08-12")
```

---

[9]"Guess_formats + R + Lubridate," Stack Overflow, accessed on January 22, 2021, https://stackoverflow.com/questions/26064292/guess-formats-r-lubridate.

Bring in the parse_date_time function:

```
bk.simple.parsed.dates <-
  parse_date_time(bk.dates.in.different.formats,
  c("Ymd", "mdY"))
bk.simple.parsed.dates
```

```
## [1] "2004-04-06 UTC" "2004-04-06 UTC" "2004-04-06 UTC" "2004-04-07 UTC"
## [5] "2004-04-06 UTC" "2004-04-07 UTC" "2014-06-28 UTC" "2014-06-30 UTC"
## [9] "2014-07-12 UTC" "2014-07-29 UTC" "2014-07-29 UTC" "2014-08-12 UTC"
```

Clearly, the parse_date_time function is another one of the smartest functions in the room. We started with a mishmash of date formats and ended with consistent vector of POSIX formats.

How many elements are in the date vector?

```
bk.simple.parsed.dates
length(bk.simple.parsed.dates)
```

```
## [1] 12
```

Use class() to show that the vector contains real dates:

```
class(bk.simple.parsed.dates)
```

```
## [1] "POSIXct" "POSIXt"
```

In this example, only two formats, "Ymd" and "mdY", were used. However, there are many other ones that could be tried if you have a large variety of formats to parse.

3.12.10 Hour Function

The hour function extracts the hour from a date-time and also allows targeted change of the hour within a date-time variable:

```
bk.x <- ymd_hms(c("2009-08-07 01:01:01"))
bk.x
```

```
## [1] "2009-08-07 01:01:01 UTC"
```

```
bk.hour <- hour(bk.x)
bk.hour
```

```
## [1] 1
```

Change the hour in an existing date-time variable from 1 to 5:

```
hour(bk.x) <- 5
bk.x
```

```
## [1] "2009-08-07 05:01:01 UTC"
```

Change the hour to 25 and Lubridate does the logical thing. It bumps the date-time to the following day:

```
hour(bk.x) <- 25
bk.x
```

```
## [1] "2009-08-08 01:01:01 UTC"
```

```
hour(bk.x) > 2
```

```
## [1] FALSE
```

3.12.11 Extract Names from Date

Use the label argument to obtain the short month name:

```
bk.date <- ymd("2019-01-15")
month(bk.date, label = TRUE)
```

```
## [1] Jan
## 12 Levels: Jan < Feb < Mar < Apr < May < Jun < Jul < Aug < Sep < ... <
Dec
```

Here's an alternative way to get a month name rather than a month number. The month name abbreviations are built into R, as month.abb:

```
the.months = 1:12
names(the.months) = month.abb
the.months[month(bk.date)]
```

```
## Jan
##   1
```

Get the full-length month name using the label and abbr arguments:

```
month(bk.date, label = TRUE, abbr = FALSE)
```

```
## [1] January
## 12 Levels: January < February < March < April < May < June < ... <
December
```

When addressing months by their index number, note that the index starts at zero, not 1:

```
month(bk.date + months(0:5), label = F)
```

```
## [1] 1 2 3 4 5 6
```

Using the label = T argument, the month abbreviations are used:

```
month(bk.date + months(0:5), label = T)
```

```
## [1] Jan Feb Mar Apr May Jun
## 12 Levels: Jan < Feb < Mar < Apr < May < Jun < Jul < Aug < Sep < ...
< Dec
```

3.12.12 Parse Periods with Hour, Minute, and Second Components

According to rdocumentation.org, the ms function in Lubridate

> *Transforms a character or numeric vector into a period object with the specified number of hours, minutes, and seconds. hms() recognizes all non-numeric characters except '-' as separators ('-' is used for negative durations). After hours, minutes and seconds have been parsed, the remaining input is ignored.*[10]

---

[10]"Ms Function | R Documentation," accessed on January 23, 2021, www.rdocumentation.org/packages/lubridate/versions/1.7.9.2/topics/ms.

ms is different from the function parse_date_time shown later in this chapter of the book.

```
ms(c("09:10", "09:02", "1:10"))
```

```
## [1] "9M 10S" "9M 2S"  "1M 10S"
```

```
bk.ms.example <-  ms(c("09:10", "09:02", "1:10"))
bk.ms.example
```

```
## [1] "9M 10S" "9M 2S"  "1M 10S"
```

The length function applied to bk.ms.example has three elements rather than one.

```
length(bk.ms.example)
```

```
## [1] 3
```

```
bk.ms.example
```

```
## [1] "9M 10S" "9M 2S"  "1M 10S"
```

hm, like ms, extracts information as hours, minutes, and seconds but assumes that the first element is hours and the second element (after the colon) is minutes:

```
bk.ms.example <-  hm(c("09:10", "09:02", "11:01"))
bk.ms.example
```

```
## [1] "9H 10M 0S" "9H 2M 0S"  "11H 1M 0S"
```

hms assumes hours, minutes, and seconds:

```
hms("7 6 5")
```

```
## [1] "7H 6M 5S"
```

If one of the elements is stated at a higher value than allowed for normal time category (e.g., more than 59 minutes), then "roll" will force it to do the right thing—bump up the next level. The hms function changed 7 hours, 65 minutes, and 5 seconds to 8 hours, 5 minutes, and 5 seconds.

```
hms("7 65 5", roll = TRUE)
```

```
## [1] "8H 5M 5S"
```

3.13 Parse Date-Time: A Lubridate Workhorse

parse_date_time is a Lubridate workhorse which converts many date character representations to valid POSIXct dates (legitimate R dates).

The following are a few examples of parse_date_time's power and flexibility.

Enter two dates, one in the format of a two-digit day, a two-digit month, and a two-digit year and the other in the form of a two-digit day, a fully spelled out month, and a four-digit year. The "orders" tell Lubridate which format applies to each date. See Appendix G for a list of all formats recognized by Lubridate.

```
bk.two.dates <- parse_date_time(c("20-01-19", "20 January 2019"), orders =
c("d m y", "d B Y"))
bk.two.dates
```

```
## [1] "2019-01-20 UTC" "2019-01-20 UTC"
```

```
class(bk.two.dates)
```

```
## [1] "POSIXct" "POSIXt"
```

How many elements are in the output?

```
length(bk.two.dates)
```

```
## [1] 2
```

```
bk.two.dates
```

```
## [1] "2019-01-20 UTC" "2019-01-20 UTC"
```

Parse based on day-month-year (two digits) plus hour-minute-second:

```
bk.date <- parse_date_time("21-02-19 20-13-06",
    orders = "dmy HMS")
bk.date
```

```
## [1] "2019-02-21 20:13:06 UTC"
```

Note that it is not necessary to put a zero in front of the day:

```
bk.date <- mdy("09/5/2016")
bk.date
```

```
## [1] "2016-09-05"
```

You can enter a date in year/month/day/hour/minute/second format and then use ymd_hms to put it into standard R date form. Dates in this format conform to ISO standards:

```
bk.date2 <- "2015-11-7 10:11:05"
bk.date3 <- ymd_hms(bk.date2)
bk.date3
```

```
## [1] "2015-11-07 10:11:05 UTC"
```

3.14 Date Validation

Is an object a valid date?

```
bk.is.this.a.date <- is.Date(as.Date("2009-08-03"))
bk.is.this.a.date
```

```
## [1] TRUE
bk.is.this.a.date <-  is.Date(9999999)
bk.is.this.a.date
```

```
## [1] FALSE
```

3.14.1 Calculate Time Difference

difftime calculates the difference in days, between two dates. As a unit of measure, it is a duration in days:

```
bk.date1 <- ymd_hms("2009-06-08 01:59:59") #end date
bk.date2 <- ymd_hms("2008-04-29 12:00:00") #beginning date
bk.diff.in.dates <- difftime(bk.date1,bk.date2)
bk.diff.in.dates
```

```
##Time difference of 404.5833 days
```

difftime is not actually a date but a quantity of days:

```
bk.is.difftime <- is.Date(bk.diff.in.dates)
bk.is.difftime
```

```
## [1] FALSE
```

You can ask Lubridate whether a variable is a difftime object:

```
bk.is.a.difftime <- is.difftime(bk.diff.in.dates)
bk.is.a.difftime
```

```
## [1] TRUE
```

Determine whether an object is POSIX:

```
bk.posix.query <- is.POSIXt(now())
bk.posix.query
```

```
## [1] TRUE
```

Determine whether an object is a timespan:

```
is.timespan(as.Date("2009-08-03"))
```

```
## [1] FALSE
```

```
is.timespan(duration(second = 1))
```

```
## [1] TRUE
```

Determine whether a date is in a leap year:

```
bk.not.leap.year <- as.Date("2019-01-01")
bk.query.leap.year <- leap_year(bk.not.leap.year)
bk.query.leap.year
```

```
## [1] FALSE
```

3.14.2 Time Zones

R uses a set of standard time zone codes. All of them are available by running OlsonNames(). The following script lists a sample of 25 of them, out of a total of 594 (as of this writing):

```
my.time.zones <- OlsonNames()
length(my.time.zones)
```

```
## [1] 594
```

```
my.sample.of.time.zones <- sample_n(as.data.frame(my.time.zones), 25)
my.sample.of.time.zones #note: since this is a sample, your
                        #output may look different from mine
```

```
##                   my.time.zones
## 1                        Iceland
## 2              Australia/Canberra
## 3                     Asia/Muscat
## 4              Atlantic/Cape_Verde
## 5                  Europe/Brussels
## 6                 Indian/Mauritius
## 7                 America/Rosario
## 8                  Asia/Chungking
## 9             Australia/Queensland
## 10                  Pacific/Kosrae
## 11                   Pacific/Chuuk
## 12                   Africa/Asmera
## 13                  Pacific/Midway
## 14               America/Anguilla
## 15                   Europe/Moscow
## 16                     Africa/Lome
## 17                  Asia/Kamchatka
## 18                Antarctica/Davis
## 19              Australia/Tasmania
## 20              America/Martinique
## 21                     Pacific/Yap
## 22                   Asia/Jakarta
```

```
## 23 America/Bahia_Banderas
## 24          America/Toronto
## 25       Atlantic/Reykjavik
```

3.14.3 Shorthand Methods to Designate Date/Times

Just key in year, month, day, and time if desired:

```
bk.easy.time1 <- make_datetime(year = 1999, month = 12,
    day = 22, sec = 10)
bk.easy.time1
```

```
## [1] "1999-12-22 00:00:10 UTC"
```

Here, create two date/times with a difference of one second:

```
bk.easy.time2 <- make_datetime(year = 1999, month = 12,
    day = 22, sec = c(10, 11))
bk.easy.time2
```

```
## [1] "1999-12-22 00:00:10 UTC" "1999-12-22 00:00:11 UTC"
```

3.14.4 Work with Weeks

Calculate the week of the year:

```
bk.x <- ymd("2012-03-26")
week(bk.x)
```

```
## [1] 13
```

Set a variable to week one and then week 54:

```
bk.week <- week(bk.x) <- 1
bk.week
```

```
## [1] 1
bk.week.set <- week(bk.x) <- 54
bk.week.set
```

```
## [1] 54
bk.week.logical.question <- week(bk.x) > 3
bk.week.logical.question
```

```
## [1] FALSE
```

3.14.5 Test Interval or Date: Is It Within Another Interval?

Define an interval between two new year's days:

```
bk.int <- interval(ymd("2001-01-01"), ymd("2002-01-01"))
bk.int
```

```
## [1] 2001-01-01 UTC--2002-01-01 UTC
```

Set up an interval of six months:

```
bk.int2 <- interval(ymd("2001-06-01"), ymd("2002-01-01"))
bk.int2
```

```
## [1] 2001-06-01 UTC--2002-01-01 UTC
```

Is May 3, 2001, within the range January 1, 2001–January 1, 2002?

```
bk.within.or.without <- ymd("2001-05-03") %within% bk.int
bk.within.or.without
```

```
## [1] TRUE
```

Ask the question: is one range within another? Is June 1, 2001, through January 1, 2002, within January 1, 2001, through January 1, 2002?

```
bk.within.or.without.2 <- bk.int2 %within% bk.int
bk.within.or.without.2
```

```
## [1] TRUE
bk.within.or.without.3 <- ymd("1999-01-01") %within% bk.int   #is 1/1/1999
within 1/1/2001-1/1/2002
bk.within.or.without.3
```

```
## [1] FALSE
bk.dates <- ymd(c("2014-12-20", "2014-12-30", "2015-01-01",
    "2015-01-03"))
bk.blackouts<- c(interval(ymd("2014-12-30"), ymd("2014-12-31")),
    interval(ymd("2014-12-30"), ymd("2015-01-03")))
bk.blackouts
```

```
## [1] 2014-12-30 UTC--2014-12-31 UTC 2014-12-30 UTC--2015-01-03 UTC
```

Using four dates, ask whether each one is within the blackout date range:

```
bk.are.dates.within.blackouts <-  bk.dates %within% bk.blackouts
bk.are.dates.within.blackouts
```

```
## [1] FALSE   TRUE FALSE   TRUE
```

Test each date in turn as to whether it is in *any* of the blackout ranges:

```
bk.dates <- ymd(c("2014-12-20", "2014-12-30",
    "2015-01-01", "2015-01-03"))
bk.blackouts<- list(interval(ymd("2014-12-30"),
    ymd("2014-12-31")), interval(ymd("2014-12-30"),
    ymd("2015-01-03")))
bk.are.dates.within.blackouts <- bk.dates %within% bk.blackouts
bk.are.dates.within.blackouts
```

```
## [1] FALSE   TRUE   TRUE   TRUE
```

3.14.6 Miscellaneous Functions: Create a Specified Time Difference

The make_difftime() function provides a plethora of options to create a spread between two date-times.[11]

```
make_difftime(1)
```

```
## Time difference of 1 secs
```

---

[11]https://cran.r-project.org/web/packages/lubridate/lubridate.pdf. Accessed on January 25, 2021.

Time difference of 1 secs:

```
bk.difftime1sec <- make_difftime(1)
bk.difftime1sec
```

Time difference of 1 secs:

```
make_difftime(60)
```

Time difference of 1 mins:

```
make_difftime(3600)
```

Time difference of 1 hours

```
make_difftime(3600, units = "minute")
```

Time difference of 60 mins

Time difference of 60 mins:

```
make_difftime(second = 90)
```

Time difference of 1.5 mins

Create a time difference of 1.5 mins:

```
make_difftime(minute = 1.5)
```

Time difference of 1.5 mins

Specify a time difference by coding individual units of measure:

```
make_difftime(second = 3, minute = 1.5, hour = 2, day = 6,
    week = 1)
```

Time difference of 13.08441 days

```
make_difftime(hour = 1, minute = -60)
```

Time difference of 0 secs

```
make_difftime(day = -1)
```

Time difference of -1 days

```
make_difftime(120, day = -1, units = "minute")
```

```
## Time differences in mins
## [1]    2 -1440
```

If no minutes are specified, Lubridate assumes zero minutes:

```
bk.date <- (ymd("2019-01-12"))
minute(bk.date)
```

```
## [1] 0
```

```
minute(ymd_hms("2009-06-08 00:59:59"))
```

```
## [1] 59
```

```
minute(bk.date)  <- 5
minute(bk.date)
```

```
## [1] 5
```

```
minute(bk.date) > 3   #logical test
```

```
## [1] TRUE
```

Show the month as a digit:

```
bk.date  #show date
```

```
## [1] "2019-01-12 00:05:00 UTC"
```

```
month(bk.date)
```

```
## [1] 1
```

3.14.7 Force a Date/Time to Be in a Different Time Zone

In the following examples, the literal date/time number is the same for each instance, but the time zone is changed in the last two. As a side note, I found replacing time zones via code to be tricky. It's easy to make a mistake. You may want to do a spot-check on any new code you write involving time zones. One (of many) time zone calculators you can use is www.timeanddate.com/worldclock/converted.html.

Start with time zone code equal to "Japan":

```
bk.date <- ymd_hms("2019-01-07 00:00:01", tz = "Japan")
bk.date
```

```
## [1] "2019-01-07 00:00:01 JST"
```

Now force the time zone to "America/Chicago":

```
bk.date2 <- force_tz(bk.date, "America/Chicago")
bk.date2
```

```
## [1] "2019-01-07 00:00:01 CST"
```

Note that this change will not alter the numbers you see. It simply changes the invisible time zone tag associated with the date-time.

Validity of daylight savings time:

```
bk.date <- ymd_hms("2010-03-14 02:05:05 UTC")
bk.date
```

```
## [1] "2010-03-14 02:05:05 UTC"
```

Now attempt to roll back that time to a time that does not exist, given daylight savings time:

```
bk.date.with.no.roll.to.valid.time <- force_tz(bk.date, "America/New_York",
roll=FALSE)
bk.date.with.no.roll.to.valid.time
```

```
## [1] NA
```

Next, instruct Lubridate to do the right thing and roll back to an appropriate time (whole number) within the daylight savings time framework:

```
bk.date.with.roll.to.valid.time <- force_tz(bk.date,
  "America/New_York", roll=TRUE)
bk.date.with.roll.to.valid.time
```

```
## [1] "2010-03-14 03:00:00 EDT"
```

3.14.8 **Working with Different Time Zones in the Same Calculations**

Set up two date-times, with a difference of approximately one hour:

```
two.dates <- c("2009-08-07 00:00:01", "2009-08-07 01:02:03")
bk.date <- ymd_hms(two.dates)
bk.date
```

```
## [1] "2009-08-07 00:00:01 UTC" "2009-08-07 01:02:03 UTC"
```

UTC is universal time code, the successor to the deprecated Greenwich Mean Time. It is a place in time where you can say "I'm putting a stake in the ground right here. Every other time on earth is measured relative to here." The entire machinery of modern civilization depends on a universally acknowledged starting point for the day.

Now assign time zones to the two date-times that make up bk.date:

```
bk.assign.time.zones <- force_tzs(bk.date, tzones =
  c("America/New_York", "Europe/Amsterdam"))
bk.assign.time.zones
```

```
## [1] "2009-08-07 04:00:01 UTC" "2009-08-06 23:02:03 UTC"
```

Since the date-times were forced to be in a specific time zone, the UTC values changed.

Another example, using four date-times, with a forced time zone imposed on each, and a time zone out, which states all those date-times in terms of the "America/New_York" time zone. The libya time zone covers Tripoli, Libya.

```
bk.date.values <- ymd_hms(c("2009-08-07 00:00:01",
  "2009-08-07 01:02:03","2009-03-03 11:02:03",
  "2009-04-01 05:02:03" ))
bk.assign.time.zones.with.tzone_out <-force_tzs(bk.date.values,
  tzones = c("America/New_York", "Europe/Amsterdam", "Japan",
  "libya"), tzone_out = "America/New_York")
bk.assign.time.zones.with.tzone_out
```

```
## [1] "2009-08-07 00:00:01 EDT" "2009-08-06 19:02:03 EDT"
## [3] "2009-03-02 21:02:03 EST" "2009-03-31 23:02:03 EDT"
```

3.14.9 Eastern Daylight Savings and Other Time Examples

Working through two time zones, with the complication of eastern daylight savings time:

```
bk.x <- ymd_hms(c("2009-08-07 00:00:01", "2009-08-07 01:02:03"))
bk.y <- force_tzs(bk.x, tzones = c("America/New_York",
  "Europe/Amsterdam"))
bk.y
```

```
## [1] "2009-08-07 04:00:01 UTC" "2009-08-06 23:02:03 UTC"
```

In bk.x, two time-dates are hard-coded and then, by default, represented internally as UTC dates. In bk.y, the original two points in time, point1 and point2, which started out in UTC, are converted to the times they would be in New York and Amsterdam, IF THOSE TIMES WERE ALSO STATED IN UTC TERMS.

Notice how point1 went from New York time of 1 minute after midnight to around 4 AM. This is because eastern daylight time = UTC – 4.

Set time zone:

```
bk.x <- ymd("2012-03-26")
tz(bk.x)
```

```
## [1] "UTC"
tz(bk.x) <- "GMT"
bk.x
```

```
## [1] "2012-03-26 GMT"
tz(bk.x) <- "America/New_York"
bk.x
```

```
## [1] "2012-03-26 EDT"
```

In this example, 12 AM in New York gets converted to 4 AM GMT (Greenwich Mean Time):

```
bk.x <- ymd_hms("2009-08-07 00:00:01", tz = "America/New_York")
bk.with.time.zone <- with_tz(bk.x, "GMT")
bk.with.time.zone
```

```
## [1] "2009-08-07 04:00:01 GMT"
```

Another time zone change, resulting in a day change:

```
bk.x.singapore <- ymd_hms("2009-08-07 00:00:01",
    tz = "Singapore")
bk.x.singapore
```

```
## [1] "2009-08-07 00:00:01 +08"
bk.with.time.zone <- with_tz(bk.x.singapore, "GMT")
bk.with.time.zone
```

```
## [1] "2009-08-06 16:00:01 GMT"
```

Year modifications:

```
bk.date <- ymd("2012-03-26")
bk.year <- year(bk.date)
bk.year
```

```
## [1] 2012
```

```
Set year to 2001:
bk.year.2001 <-  year(bk.date) <- 2001
bk.year.2001
```

```
## [1] 2001
bk.year.1995 <- year(bk.date) > 1995
bk.year.1995
```

```
## [1] TRUE
```

3.14.10 Internationalization

According to a stack overflow article[12]

---

[12]"Php – List of All Locales and Their Short Codes? – Stack Overflow," accessed on January 24, 2021, https://stackoverflow.com/questions/3191664/list-of-all-locales-and-their-short-codes.

In computing, locale is a set of parameters that defines the user's language, country and any special variant preferences that the user wants to see in their user interface. Usually a locale identifier consists of at least a language identifier and a region identifier.

To see where your system thinks you live, enter the following:

```
Sys.getlocale("LC_TIME") #your response may vary slightly

## [1] "English_United States.1252"
```

Locales are specified as an ISO standard. Take note that locales are not defined with perfection. I attempted to use the locale "ro_RO.utf8" (Romania) and got the message "Error in Sys.setlocale("LC_TIME", locale):(converted from warning) OS reports request to set locale to "ro_RO.utf8" cannot be honored."

So, like any author with plenty of things to do, I moved on and selected a new locale, in Germany:

```
bk.date1 <-  "Ma 2012 august 14 11:28:30 "
bk.german.date <- ymd_hms(bk.date1, locale = "German")
bk.german.date

## [1] "2012-08-14 11:28:30 UTC"
```

Note   On Mac, your results may vary from the Windows version of R.

Next, I use my locale value to show a Texas, USA, date:

```
bk.date.new <-  "Ma 2012 august 14 11:28:30 "
bk.texas.date <- ymd_hms(bk.date.new,
  locale = "English_United States.1252")
bk.texas.date

## [1] "2012-08-14 11:28:30 UTC"
```

3.14.10.1 Is There Any Good Locale News?

Even after discussing the matter with experts on stackoverflow.com, it is not always clear where to get good locale names, at least for the Windows OS. Richie Cotton suggested the following site for locale names: https://docs.microsoft.com/en-us/windows/desktop/Intl/language-identifier-constants-and-strings. I used this Microsoft website to obtain the locale for Belarus. The site shows "LANG_LANG_BELARUSIAN" as the locale name. That did not work. However, I tried "BELARUSIAN" without the "LANG_" prefix and the code works fine.

```
bk.date1 <- "Ma 2012 august 14 11:28:30 "
ymd_hms(bk.date1, locale = "LANG_LANG_BELARUSIAN")

##Error in Sys.setlocale("LC_TIME", locale) :
##  (converted from warning) OS reports request to set locale to ##"LANG_
LANG_BELARUSIAN" cannot be honored
##In addition: Warning message:
##In Sys.setlocale("LC_TIME", locale) :
##  OS reports request to set locale to "LANG_LANG_BELARUSIAN"
##cannot be honored

ymd_hms(bk.date1,
  locale = "BELARUSIAN") # as modified by guessing
[1] "2012-08-14 11:28:30 UTC"
```

3.14.11 now( ), Rollback, and Rounding

3.14.11.1 Right Now

Current date and time, based on your system setting:

```
bk.current.date.time <- now()
bk.current.date.time

## [1] "2021-01-12 17:11:12 CST"
```

now() with time zone:

```
bk.current.date.time <- now(tzone = "Pacific/Midway")
bk.current.date.time
```

```
## [1] "2021-01-12 12:11:12 SST"
```

I'm not sure how often you would use it, but if you want to see the date which R uses to start its own internal time, just type in "origin":

```
origin
```

```
## [1] "1970-01-01 UTC"
```

For just the current date, omitting the hour and minute:

```
bk.today <- today()
bk.today
```

```
## [1] "2021-01-12"
```

```
bk.today.greenwich.time <- today("GMT")
bk.today.greenwich.time
```

```
## [1] "2021-01-12"
```

```
bk.today.have.not.had.dinner.on.mars.yet <-
  as.Date("2035-01-01")
bk.true.or.false <- today() <
    bk.today.have.not.had.dinner.on.mars.yet
bk.true.or.false
```

```
## [1] TRUE
```

3.14.11.2 Rollback

Rollback changes a date to the last day of the previous month or the first day of the month:

```
bk.date <- ymd("2019-05-20")
bk.rolled.back <- rollback(bk.date)
bk.rolled.back
```

```
## [1] "2019-04-30"

bk.dates <- bk.date + months(0:2)
bk.dates
## [1] "2019-05-20" "2019-06-20" "2019-07-20"

bk.multiple.dates <- rollback(bk.dates)
bk.multiple.dates

## [1] "2019-04-30" "2019-05-31" "2019-06-30"

bk.date <- ymd_hms("2010-03-03 12:44:22")
bk.date.rolled.back <- rollback(bk.date)
bk.date.rolled.back

## [1] "2010-02-28 12:44:22 UTC"
```

In this syntax, Lubridate rolls the day back to the first of the month:

```
bk.roll.to.first <- rollback(bk.date, roll_to_first = TRUE)
bk.roll.to.first

## [1] "2010-03-01 12:44:22 UTC"
```

Ignore hours, minutes, and seconds:

```
bk.roll.back.drop.hours.minutes.seconds <- rollback(bk.date, preserve_hms =
FALSE)
bk.roll.back.drop.hours.minutes.seconds

## [1] "2010-02-28 UTC"

bk.roll.back.no.keep.hours.minutes.seconds <- rollback(bk.date,
  roll_to_first = TRUE, preserve_hms = FALSE)
bk.roll.back.no.keep.hours.minutes.seconds

## [1] "2010-03-01 UTC"
```

3.14.11.3 Rounding

The rounding function is intuitive. It includes floor, ceiling, traditional rounding, week start, time unit, and sync to the next "boundary." The following examples are from the Lubridate manual, with some annotations[13]:

Almost but not quite 2AM:

```
date <- as.POSIXct("2009-03-08 01:59:59") # DST boundary
```

Print fractional seconds:

```
options(digits.secs=6)
x <- ymd_hms("2009-08-03 12:01:59.23")

round_date(x, ".5s")

## [1] "2009-08-03 12:01:59 UTC"
round_date(x, "sec")

## [1] "2009-08-03 12:01:59 UTC"
round_date(x, "second")

## [1] "2009-08-03 12:01:59 UTC"
round_date(x, "minute")

## [1] "2009-08-03 12:02:00 UTC"
round_date(x, "5 mins")

## [1] "2009-08-03 12:00:00 UTC"
round_date(x, "hour")

## [1] "2009-08-03 12:00:00 UTC"
round_date(x, "2 hours")

## [1] "2009-08-03 12:00:00 UTC"
round_date(x, "day")

## [1] "2009-08-04 UTC"
round_date(x, "week")
```

---

[13]https://cran.r-project.org/web/packages/lubridate/lubridate.pdf. Accessed on January 24, 2021.

```
## [1] "2009-08-02 UTC"
round_date(x, "month")
```

```
## [1] "2009-08-01 UTC"
round_date(x, "bimonth")
```

```
## [1] "2009-09-01 UTC"
round_date(x, "quarter") == round_date(x, "3 months")
```

```
## [1] TRUE
round_date(x, "halfyear")
```

```
## [1] "2009-07-01 UTC"
round_date(x, "year")
```

```
## [1] "2010-01-01 UTC"
x <- ymd_hms("2009-08-03 12:01:59.23")
floor_date(x, ".1s")
```

```
## [1] "2009-08-03 12:01:59.2 UTC"
floor_date(x, "second")
```

```
## [1] "2009-08-03 12:01:59 UTC"
floor_date(x, "minute")
```

```
## [1] "2009-08-03 12:01:00 UTC"
floor_date(x, "hour")
```

```
## [1] "2009-08-03 12:00:00 UTC"
floor_date(x, "day")
```

```
## [1] "2009-08-03 UTC"
floor_date(x, "week")
```

```
## [1] "2009-08-02 UTC"
floor_date(x, "month")
```

```
## [1] "2009-08-01 UTC"
floor_date(x, "bimonth")
```

```
## [1] "2009-07-01 UTC"
floor_date(x, "quarter")
```

```
## [1] "2009-07-01 UTC"
floor_date(x, "season")
```

```
## [1] "2009-06-01 UTC"
floor_date(x, "halfyear")
```

```
## [1] "2009-07-01 UTC"
floor_date(x, "year")
```

```
## [1] "2009-01-01 UTC"
x <- ymd_hms("2009-08-03 12:01:59.23")
ceiling_date(x, ".1 sec")
```

```
## [1] "2009-08-03 12:01:59.2 UTC"
ceiling_date(x, "second")
```

```
## [1] "2009-08-03 12:02:00 UTC"
ceiling_date(x, "minute")
```

```
## [1] "2009-08-03 12:02:00 UTC"
ceiling_date(x, "5 mins")
```

```
## [1] "2009-08-03 12:05:00 UTC"
ceiling_date(x, "hour")
```

```
## [1] "2009-08-03 13:00:00 UTC"
ceiling_date(x, "day")
```

```
## [1] "2009-08-04 UTC"
ceiling_date(x, "week")
```

```
## [1] "2009-08-09 UTC"
ceiling_date(x, "month")
```

```
## [1] "2009-09-01 UTC"
ceiling_date(x, "bimonth") == ceiling_date(x, "2 months")
```

```
## [1] TRUE
ceiling_date(x, "quarter")
```

```
## [1] "2009-10-01 UTC"
ceiling_date(x, "season")
```

```
## [1] "2009-09-01 UTC"
ceiling_date(x, "halfyear")
```

```
## [1] "2010-01-01 UTC"
ceiling_date(x, "year")
```

```
## [1] "2010-01-01 UTC"
x <- ymd("2000-01-01")
ceiling_date(x, "month")
```

```
## [1] "2000-02-01"
ceiling_date(x, "month", change_on_boundary = TRUE)
```

```
## [1] "2000-02-01"
```

Seconds are treated similarly to other time measures. Note that seconds, unlike higher orders of time measures such as minutes or hours, are invariant. In other words, a second during a leap year is the same duration as a second in a non-leap year.

```
bk.date <- ymd("2012-03-26")
second(bk.date)
```

```
## [1] 0
```

Set number of seconds to 1:

```
second(bk.date) <- 1
bk.date
```

```
## [1] "2012-03-26 00:00:01 UTC"
```

If a date-time is set to 61 seconds, then its value will become 1 second:

```
second(bk.date) <- 61
bk.date
```

```
## [1] "2012-03-26 00:01:01 UTC"
second(bk.date) > 2
```

```
## [1] FALSE
## ** truncated time-dates **
```

```
bk.x <- c("2011-12-31 12:59:59", "2010-01-01 12:11",
    "2010-01-01 12", "2010-01-01")
bk.x.truncated <- ymd_hms(bk.x, truncated = 3)
bk.x.truncated
```

```
## [1] "2011-12-31 12:59:59 UTC" "2010-01-01 12:11:00 UTC"
## [3] "2010-01-01 12:00:00 UTC" "2010-01-01 00:00:00 UTC"
```

```
bk.x <- c("2011-12-31 12:59", "2010-01-01 12",
  "2010-01-01")
```

In this syntax, note the absence of seconds in the output:

```
bk.x.truncated.2 <- ymd_hm(bk.x, truncated = 2)
bk.x.truncated.2
```

```
## [1] "2011-12-31 12:59:00 UTC" "2010-01-01 12:00:00 UTC"
## [3] "2010-01-01 00:00:00 UTC"
```

3.14.11.4 Automatic Roll over Arithmetic

Since months do not all have the same number of days, Lubridate cannot use simple math to calculate, for example, the date a month from now. For example, if someone says, "add one month to January 31," he is typically thinking of February 28, not March 3. Lubridate adds or subtracts so that the end result does not exceed the last day of the newly calculated month.

This code does not give you the result you expect:

```
bk.jan <- ymd_hms("2010-01-31 03:04:05")
bk.three.months.out <- bk.jan + months(1:3)
bk.three.months.out
```

```
## [1] NA                      "2010-03-31 03:04:05 UTC"
## [3] NA
```

Now use our shiny new toy, %m+%. There is no rollover. Lubridate creates three month-time values corresponding to real and expected values when moving forward three months.

```
bk.jan.do.the.right.thing <- bk.jan %m+% months(1:3)
bk.jan.do.the.right.thing
```

```
## [1] "2010-02-28 03:04:05 UTC" "2010-03-31 03:04:05 UTC"
## [3] "2010-04-30 03:04:05 UTC"
leap <- ymd("2012-02-29")
leap %m+% years(1)   #jumps to last day in February
```

```
## [1] "2013-02-28"
leap %m+% years(-1) #same date, previous year
```

```
## [1] "2011-02-28"
leap %m-% years(1)   # the minus sign means it goes back a year
```

```
## [1] "2011-02-28"
```

Q: Why couldn't young Billy-Bob try out regular expressions at home?

A: His mom wouldn't let him play with matches.

## CHAPTER 4

Regular Expressions: Introduction

First impressions of regular expressions are rarely positive. They look arcane, a throwback to the early decades of modern computing, when GUIs and object-oriented programming were a distant future. However, once you get past its look and feel, regular expressions give you serious power to search, filter, and manipulate text and numbers with speed and minimal code.

Once you reach a foundational understanding, regular expressions are often quicker to write than their equivalent R native code. You don't need to remember the specific commands and package required. In some cases, RegEx can elegantly execute one line of code in lieu of ten lines of traditional logic. If you use multiple languages in your work, it's good to know that regular expressions are part of many languages such as Java, Python, JavaScript, and .NET. Another benefit—RegEx is stable over time. Any changes or additions do not alter past functionality.

Knowing regular expressions will take you to the next level of expertise. It shows others that you can focus, learn complex syntax, and proactively acquire powerful tools. This book's examples conform to R syntax, permitting you to do complex searches (even across multiple lines), replacements, and validations. You'll be at the RegEx beginner level quickly, able to create simple patterns. Practice for a week and you've got a lifetime skill as part of your core competence.

Besides programming languages, RegEx is often included with text editors and some applications. For example, the find feature on Notepad++ has an option to select regular expression queries or even extended RegEx queries. Some applications, such as Grouper by BIS, require the use of regular expressions to be fully functional. RegEx also shines when running against large datasets.

In practice, you'll choose RegEx or other R functions within Stringr, stringi, and so on as needed. The Internet is replete with examples of regular expressions for typical business/data science tasks, such as validation of American Express cards or Social Security numbers. Unless you just enjoy developing code from scratch, copying existing code from the Internet is often the shortest path to a solution. As you grow in RegEx skills, creating a simple statement will be the easier choice. One thing you can be sure of—*every line of RegEx code in this book runs in R.*

When perusing the Internet looking for RegEx examples, be aware that many coding examples don't bother to disclose what language they are written in. Some may not work in R without modification.

4.1 RegEx: A Few Tips to Get Started

If you are like me, you use the Internet to find examples of code. Keep in mind that many of the regular expressions you will see are not R-specific. For R, you'll need to use *two* backslashes rather than one.

As copied from the Internet:[1] `^[\w\.=-]+@[\w\.-]+\.[\w]{2,3}$`

As coded in R: `^[\\w\\.=-]+@[\\w\\.-]+\\.[\\w]{2,3}$`

Load the following libraries to ensure any code from the book works:

- library(tidyverse)
- library(stringr)
- library(stringi)
- library(readr)[2]
- library(tidyr)

---

[1] "Regular Expression Library," accessed on March 14, 2020, `http://regexlib.com/Search.aspx?k=email&c=-1&m=-1&ps=20`.

[2] "Read Rectangular Text Data," accessed on April 7, 2020, `https://readr.tidyverse.org/`.

4.2 Challenges and Promises of Regular Expressions[3]

One's first impression of regular expressions is stupefaction. Regular expressions are

- Cryptic, with seemingly random choices for syntax.

- Dense, with much information contained in a small number of characters.

- Complex. Sometimes crafting RegEx feels like you are trying to thread a sewing machine while it is running.

- Packed with double-duty symbols, depending on context.

- Dialect dependent. There are many dialects, including R, Java, .NET, Python, Perl, PHP, and PCRE, to name a few (see Appendix G). When searching the Internet for a particular RegEx solution, keep this in mind. There is no "universal" RegEx.

Fortunately, they are also

- Interchangeable with a given language or environment. Once you have validated your RegEx, it can often be used in many places.

- Supported by most languages. Tweaking between flavors is required, but the overall structure of the RegEx will remain largely the same.

- Clean, with little ambiguity. Allows your code to be cleaner. You can see the logic on one line.

---

[3]"Why Are Regular Expressions Difficult?", accessed on March 27, 2020, www.johndcook.com/blog/2019/06/19/why-regex/.

- Faster to run and write compared to the alternatives of various IF, IFELSE, and other statements.

- Sometimes able to enable simpler coding methods. They enable complex logic with multiple conditions in searches, validations, and text/numeric replacements.

I know one company, BIS, based out of Oklahoma City, with an AI-based product, Grouper, for converting unstructured data (e.g., an invoice is scanned, PDF format) to structured data for analysis. Regular expressions enable selection of certain documents in batches of hundreds of thousands of invoices. The English equivalent might be: find any occurrence of "invoice:", look for a numeric amount following a dollar sign, find a second amount if the invoice is split up, ignore if you see the word "duplicate" anywhere, and so on.

CHAPTER 5

Typical Uses

5.1 Test for a Match[1]

The simplest patterns match exact strings:

```
x <- c("apple", "banana", "pear")
str_extract(x, "an")
```

```
## [1] NA    "an" NA
```

This syntax is case sensitive:

```
bananas <- c("banana", "Banana", "BANANA")
str_detect(bananas, "banana")
```

```
## [1]  TRUE FALSE FALSE
```

```
str_detect(bananas, regex("banana", ignore_case = TRUE))
```

```
## [1] TRUE TRUE TRUE
```

The next step up in complexity is ., which matches any character except a newline:

```
str_extract(x, ".a.")
```

```
## [1] NA    "ban" "ear"
```

---

[1]"Regular Expressions," accessed on March 25, 2020, https://cran.r-project.org/web/packages/stringr/vignettes/regular-expressions.html.

© William Yarberry 2021
W. Yarberry, *CRAN Recipes*, https://doi.org/10.1007/978-1-4842-6876-6_5

You can allow . to match everything, including \n, by setting dotall = TRUE:

```
str_detect("\nX\n", ".X.")

## [1] FALSE

str_detect("\nX\n", regex(".X.", dotall = TRUE))
## [1] TRUE
```

5.2 Validation (e.g., Passwords)[2]

grepl outputs a TRUE or FALSE based on the match. This could be used in an application to ensure the user provides a sufficiently robust password.

```
is.validpw <- function(x) {
  grepl('(?x)                  # free-spacing mode
          ^                    # assert position = beginning of string
            (?=.*[[:upper:]])  # match one uppercase letter
            (?=.*[[:lower:]])  # match one lowercase letter
            (?=.*[[:punct:]])  # match one special character
            [ -~]{8,32}        # match printable ascii characters
          $                    # assert position = end of string', x, perl =
                                 TRUE)
}

is.validpw(c('#Password', '_PASSWORD', 'Password', 'password',
             'passwor', 'pAsswords', 'Pa&sword', 'Pa&s word'))

## [1]   TRUE FALSE FALSE FALSE FALSE FALSE  TRUE  TRUE
```

---

[2]"Validating Password in R with RegEx and POSIX Character Classes," Stack Overflow, accessed on March 25, 2020, https://stackoverflow.com/questions/40898055/validating-password-in-r-with-regex-and-posix-character-classes.

5.3 Find All Numbers and Periods in a String

```
test <- c("desk#82", "chair weight $33.20")
my.regex <- "[^0-9\\.]"
gsub(my.regex, "", test)

## [1] "82"      "33.20"
```

5.4 Change Characters[3]

Add to "a" or a cluster of a's with a "z" before the "a" and a "z" after the last "a" in the cluster:

```
sub("(a+)", "z\\1z", c("abc", "def", "cba a", "aa"), perl=TRUE)

## [1] "zazbc"    "def"       "cbzaz a" "zaaz"

gsub("(a+)", "z\\1z", c("abc", "def", "cba a", "aa"), perl=TRUE)

## [1] "zazbc"       "def"          "cbzaz zaz" "zaaz"
```

5.5 Format Strings

Use a regular expression to change a number with periods separating the thousands into commas:

```
string1 <- "234.456.100"
gsub("\\.", ",", string1)

## [1] "234,456,100"
```

---

[3]"Regular Expressions with grep, Regexp and sub in the R Language," accessed on March 25, 2020, www.regular-expressions.info/rlanguage.html.

5.6 Email

There are many variations of RegEx email validation. In the following example, my.regex is the pattern and my.emails contains a vector of four strings. The R function str_match outputs the first three elements of the vector because they are valid, but the last element, "not.an.email", is not valid, presenting as "NA." Some people, with more patience than me, have developed better ones than this. Many large RegExes for email validation can be ten lines or more. The one shown as follows is adequate for most people[4]:

```
my.regex <- "^\\w+[\\w-\\.]*\\@\\w+((-\\w+)|(\\w*))\\.[a-z]{2,3}$"
my.emails <- c("john.smith@anything.com","___fred@x.com","123Sally@Y.
net","not.an.email")
str_match (my.emails, my.regex)
```

```
##        [,1]                         [,2] [,3] [,4]
## [1,] "john.smith@anything.com" ""   NA   ""
## [2,] "___fred@x.com"           ""   NA   ""
## [3,] "123Sally@Y.net"          ""   NA   ""
## [4,] NA                        NA   NA   NA
```

Here's how to read the RegEx, going from left to right:

- ^ – What comes next refers to the start of the string.

- \\w – Matches any single character classified as a "word" character (alphanumeric or "-").

- + – Match one or more times.

- \\w.

- - – Dash.

- \\. – Dot.

- * – Match zero or more times.

- @ – Needs to have an @ sign.

- \\w.

---

[4]"Regular Expression Library," accessed on March 15, 2020, http://regexlib.com/Search.aspx?k=email&c=-1&m=-1&ps=20.

- +.

- [– Starting bracket indicates the start of a "character class," which means "any character from a, b, or c" (a character class may use ranges, e.g., [a-d] = [abcd]).

- - – Dash.

- \\w.

- * – Match zero or more times.

- \\. – Dot.

- [a-z] – a through z.

- {2,3} – Combined with the following $ sign, indicates that the end of the email needs to be two or three characters.

5.7 Example Validations with RegEx

5.7.1 Amex Card Number

Validate the configuration of an American Express credit card number. The test.amex. bad string has a second digit which is not a 4 or 7 and therefore tests as false (bad Amex card number):

```
test.amex.good <-      "348282246310005"
test.amex.bad <-       "381449635398431"

grepl("^3[47][0-9]{13}$",test.amex.good)

[1] TRUE
grepl("^3[47][0-9]{13}$",test.amex.bad)

[1] FALSE
```

Here is an explanation of mask for an Amex number:

^ – Beginning of the string (Amex number in this case).

3 – Number must start with a 3.

[47] – The character string must include either a 4 *or* a 7 (not a range).

[0-9] – Includes all single-digit numbers.

{13} – How many of the preceding characters are to be found in the target search.

$ – End of the string.

5.7.2 Email

Here's an email validation example (one of many possible patterns). For the string test. bad.email, I intentionally keyed in @ twice, to show how a false condition is generated— false, in this case, meaning that the pattern in the test.bad.email string was not matched to the RegEx:

```
pattern1 <- "^[[:alnum:].-_]+@[[:alnum:].-]+$"
test.good.email <-  "byarberry@iccmconsulting.net"
test.bad.email  <-  "mr-got-in-a-hurry@@bargain.com"
grepl(pattern1,test.good.email)

## [1] TRUE
grepl("pattern1",test.bad.email)

## [1] FALSE
```

5.7.3 IP4 Address

```
pattern1 <- "\\b(25[0-5]|2[0-4][0-9]|1[0-9][0-9]|[1-9]?[0-9])\\.
(25[0-5]|2[0-4][0-9]|1[0-9][0-9]|[1-9]?[0-9])\\.(25[0-5]|2[0-4][0-9]|1[0-9]
[0-9]|[1-9]?[0-9])\\.(25[0-5]|2[0-4][0-9]|1[0-9][0-9]|[1-9]?[0-9])\\b"

test.ip4.address.good <- "172.16.254.1"
test.ip4.address.bad <- "1723.16.254.2"
grepl(pattern1, test.ip4.address.good)
```

```
[1] TRUE
```

```
grepl(pattern1, test.ip4.address.bad)
```

```
[1] FALSE
```

5.7.4 US Social Security Number[5]

Figure 5-1 shows an example of a US Social Security card, as presented on ssa.gov.

Figure 5-1. *Example of a Social Security card.*

```
pattern1 = "\\b(?!000)(?!666)[0-8][0-9]{2}[- ](?!00)[0-9]{2}[-
  ](?!0000)[0-9]{4}\\b"
test.ssn.good <- "419-75-4971"
test.ssn.bad <- "418-833-4971"
grepl(pattern1, test.ssn.good, perl = TRUE)
```

```
## [1] TRUE
grepl(pattern1, test.ssn.bad, perl = TRUE)
```

```
## [1] FALSE
```

---

[5]"Social Security History," accessed on March 24, 2020, www.ssa.gov/history/reports/ssnreportex.html.

I initially got an error message from R, stating that the pattern1 regular expression was invalid. It being late at night as I'm writing this book, I decided to stick **"perl = TRUE"** at the end of the grepl function. It worked. There are many flavors of regular expression in the world such as Perl, Java, POSIX 1003.2, C++, ASP.NET, and many others. R supports POSIX 1003.2 (the default) and Perl. Just be aware that if you find an example RegEx pattern somewhere on the Internet, it might not work, unless it falls within the two standards supported by R.

When finding solutions from other languages, try replacing "\" with "\\" if you get an error message.

5.8 Remove Path Information from a String, Showing Only File Name

```
file.name <- "C:\budget\\second.quarter.xlsx"
y <- gsub(pattern = "C:\budget",replacement = "", x = file.name)
z <- sub(pattern = "^.{1}",replacement = "",x = y)
z

##[1] "second.quarter.xlsx"
```

In the preceding presentation, a second slash is required after budget in the file name.

5.9 Remove Non-digits

The expression my.regex means that only 0,1,2,3,4,5,6,7,8,9 will match x. In this example, the period in the pie value will be removed:

```
x <- "The corona virus is exceedingly worrisome 3.14159"
my.regex <- "[^0-9]+"
gsub(my.regex, "",x)

## [1] "314159"
```

5.10 Extract File Type from a URL

```
my.URL <- "http://www.retro51.com/_notes/Disney_Booklet_sm.pdf"
tools::file_ext(sub("\\?.+", "", my.URL))
## [1] "pdf"
```

5.10.1 Replace Any String with Same, Adjacent Letters

```
x <- c("appleeee","corn","ccccheze", "food")
gsub("\\s*\\b(?=\\w*(\\w)\\1)\\w+\\b", x, replacement = " ", perl = TRUE)
## [1] " "     "corn" " "     " "
```

5.10.2 Find Adjacent Duplicates but Not Repeats of Three or More

```
x <- "cat cat dog frog hog horse horse horse cat"
my.regex <- "\\b(\\w+)\\s+\\1\\b"
gsub(my.regex,x,replacement = " ", perl = TRUE)

## [1] "  dog frog hog   horse cat"
```

5.11 Some Other RegEx Uses[6]

- Reversing a string

- Matching html elements (href's and img's anchors)

- Working with common log format files

- Calculating basic statistics and analytics of character data

---

[6]Gaston Sanchez, Handling Strings with R, accessed on April 7, 2020, http://gastonsanchez.com/r4strings/.

CHAPTER 6

Some Simple Patterns

Regular expressions, taken to their extreme, can make you feel irregular. For many R programming tasks, only a few meta-characters need be used. Table 6-1 is from an excellent Loyola Marymount University website.[1] In R, the regular expression will be enclosed in quotes and used in one of the seven functions listed in Chapter 9, "The Magnificent Seven."

Table 6-1. *Example regular expressions*

Example regular expression	Matches any string that
Hello	Contains {hello}.
gray|grey	Contains {gray, grey}.
gr(a|e)y	Contains {gray, grey}.
gr[ae]y	Contains {gray, grey}.
b[aeiou]bble	Contains {babble, bebble, bibble, bobble, bubble}.
[b-chm-pP]at|ot	Contains {bat, cat, hat, mat, nat, oat, pat, Pat, ot}.
colo?r	Contains {color, colour}.
rege(x(es)?|xps?)	Contains {reqex, reqexes, reqexp, reqexps}.
go*gle	Contains {qqle, qoqle, qooqle, qoooqle, qooooqle,…}.
go+gle	Contains {qoqle, qooqle, qoooqle, goooogle,…}.
g(oog)+le	Contains {qooqle, qooqooqle, qooqooqooqle, qoogoogoogoogle,…}.

(continued)

---

[1]"Regular Expressions," accessed February on 4, 2020, https://cs.lmu.edu/~ray/notes/regex/.

© William Yarberry 2021
W. Yarberry, *CRAN Recipes*, https://doi.org/10.1007/978-1-4842-6876-6_6

Table 6-1. (*continued*)

Example regular expression	Matches any string that
z{3}	Contains {zzz}.
z{3,6}	Contains {zzz, zzzz, zzzzz, zzzzzz}.
z{3,}	Contains {zzz, zzzz, zzzzz,...}.
[Bb]rainf**k	Contains {Brainf**k, brainf**k>.
\d	Contains {0,1,2,3,4,5,6,7,8,9}.
\d{5}(-\d{4})?	Contains a US zip code.
1\d{10}	Contains an 11-digit string starting with a 1.
[2-9]|[12]\d|3[0-6]	Contains an integer in the range 2-36 inclusive.
Hello\nworld	Contains Hello followed by a newline followed by world.
mi.....ft	Contains a nine-character (sub)string beginning with mi and ending with ft. (Note: Depending on context, the dot stands either for "any character at all" or "any character except a newline.") Each dot is allowed to match a different character, so both Microsoft and Minecraft will match.
\d+(\.\d\d)?	Contains a positive integer or a floating-point number with exactly two characters after the decimal point.
[^i*&2@]	Contains any character other than an i, asterisk, ampersand, 2, or at sign.
//[^\r'\n]*[\r\n]	Contains a Java or C# slash-slash comment.
^dog	Begins with "dog."
dog$	Ends with "dog."
^dog$	Is exactly "dog."

6.1 Extract Lowest-Level Subdirectory from a File Path

This handy routine identifies the subdirectory containing a file(s)[2]:

```
x <- "C:/temp/budget/spreadsheet.xlsx"
gsub(".*/(.*)/[^/]+$","\\1",x)

## [1] "budget"
```

As always in R, there are multiple solutions to this task. RHertel suggests the following non-RegEx approach[3]:

```
x <- "C:/temp/budget/spreadsheet.xlsx"
basename(dirname(x))

## [1] "budget"
```

6.2 Find URL in a String of Text[4]

```
text1 <- c("s@1212a www.abcd.com www.cats.com",
           "www.boo.com",
           "asdf",
           "blargwww.test.comasdf")

pattern <- "www\\..*?\\.com"
```

Get information about where the pattern matches text1 and then extract:

```
m <- gregexpr(pattern, text1)
regmatches(text1, m)
```

---

[2]"RegEx to Extract String from a File Path R?", Stack Overflow, accessed on March 22, 2020, https://stackoverflow.com/questions/51463499/regex-to-extract-string-from-a-file-path-r.

[3]https://stackoverflow.com/questions/51463499/regex-to-extract-string-from-a-file-path-r. Accessed on January 21, 2021.

[4]"Extract Websites Links from a Text in R," Stack Overflow, accessed on March 22, 2020, https://stackoverflow.com/questions/15579119/extract-websites-links-from-a-text-in-r.

```
## [[1]]
## [1] "www.abcd.com" "www.cats.com"
##
## [[2]]
## [1] "www.boo.com"
##
## [[3]]
## character(0)
##
## [[4]]
## [1] "www.test.com"
```

Regular expressions take some practice. If you type "?RegEx" (without the quotes) in RStudio, you'll get the following helpful information, shown in Figure 6-1.

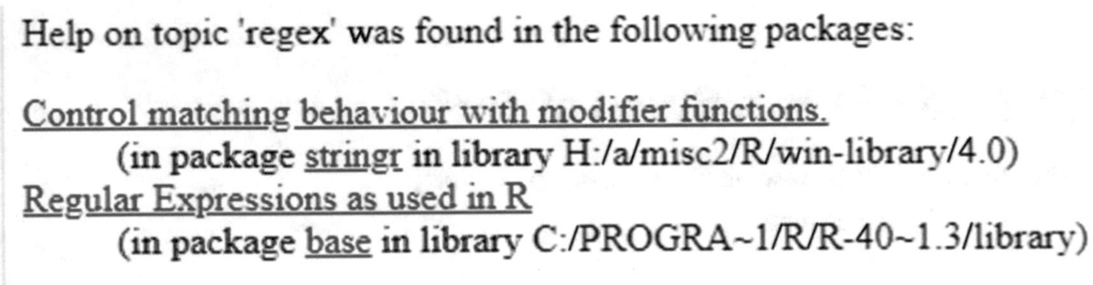

Help on topic 'regex' was found in the following packages:

Control matching behaviour with modifier functions.
    (in package stringr in library H:/a/misc2/R/win-library/4.0)
Regular Expressions as used in R
    (in package base in library C:/PROGRA~1/R/R-40~1.3/library)

Figure 6-1. *Help on regular expressions obtained by typing ?RegEx in console*

6.3 Find Zip Codes Within a String (e.g., Full Addresses)[5]

Note again the use of perl = TRUE to ensure this code works correctly in R. It is not always necessary, but most of the time, the specification is a pretty good bet.

```
zips <- data.frame(id = seq(1, 5),
 address = c("Company, 18540 Main Ave., City, ST 12345",
 "Company 18540 Main Ave. City ST 12345-0000",
 "Company 18540 Main Ave. City State 12345",
 "Company, 18540 Main Ave., City, ST 12345 USA",
 "Company, One Main Ave Suite 18540, City, ST 12345"))
regmatches(zips$address,
 gregexpr ('[0-9]{5}(-[0-9]{4})?(?!.*[0-9]{5}(-[0-9]{4})?)',
 zips$address, perl = TRUE))

## [[1]]
## [1] "12345"
##
## [[2]]
## [1] "12345-0000"
##
## [[3]]
## [1] "12345"
##
## [[4]]
## [1] "12345"
##
## [[5]]
## [1] "12345"
```

---

[5]"R – RegEx to Extract US Zip Codes but Not Faux Codes," Stack Overflow, accessed on March 22, 2020, https://stackoverflow.com/questions/25180752/regex-to-extract-us-zip-codes-but-not-faux-codes.

6.4 Codes for Regular Expressions

Meta-characters are characters which have assigned meanings, such as "this is a digit" or "this means the characters must be at the end of a string." They define search criteria and text manipulations. Die-hard RegEx fans wear T-shirts saying "I never metacharacter I didn't like."

Meta-characters include

( ) [ ] { } ^ $. \ ? * + | \d

If you want to treat a character normally functioning as a meta-character, then put a slash before it (escape character). The previous section showed example meta-characters and notations. Tables 6-2 and 6-3 show meta- and repeat characters, respectively.[6]

Table 6-2. *Meta-characters*

Character	Meaning
^	Beginning of string. Note: Within brackets, such as [^a-k], it means negation or everything except a through k.
$	End of string.
.	Any character except newline.
*	Match zero or more times.
+	Match one or more times.
?	Match zero or one time, or shortest match.
|	Alternative.
()	Grouping, "storing."
[]	Set of characters.
{}	Repetition modifier.
\	Quote or special.

---

[6]"Regular Expressions in Perl – a Summary with Examples," accessed on March 13, 2020, `http://jkorpela.fi/perl/regexp.html`.

Table 6-3. *Repeat characters*

a*	Zero or more a's.
a+	One or more a's.
a?	Zero or one a (i.e., optional a).
a{m}	Exactly m a's.
a {m,}	At least m a's.
a{m,n}	At least m but at most n a's.
Repetition?	Same as repetition but the *shortest* match is taken.

CHAPTER 7

Character Classes

A character class is how RegEx understands which characters should be considered for a match or "anti-match"—anything but what is shown. Note that "class" in this context has nothing to do with the R statement "class(x)," which gives you the class of object x.

Character classes start out simple. Just enclose match characters in brackets. [fk] means select anything containing a "f" or "k." It does not matter which one is first, and you can have as many letters, numbers, or anything else you need.

A hyphen between two characters indicates a range. The range should always be from low to high (a to z, not z to a), for example, [0-9] for all digits and [a-z] for lowercase letters. A character class matches only a *single* character. For example, "gr[ae]y" does not match graay or graey. *Remember that "ae" means either a or e but not both.*

Negated character class: If you type a caret *after* the opening square bracket, the pattern matching is based on any character, not in the character class. Line break characters need to be included in the expression to ensure accurate results. For example, [^0-7\r\n] matches any character that is not zero through seven and not a line break. Another example is t[^i] which means the pattern is a t followed by a single character which is not an i.

7.1 DOT

Dot means "match any character." The only point of uncertainty is whether or not to include the line break or not. Various flavors of RegEx treat it differently. In R, the inclusion of "dotall = TRUE" ensures that line breaks (\n) will be included. In the following example,[1] the first case shows X surrounded by line breaks. Without dotall = TRUE, the match is rejected. In the second case, line breaks are included, so a match is shown.

---

[1]"Regular Expressions," accessed on March 14, 2020, https://cran.r-project.org/web/packages/stringr/vignettes/regular-expressions.html.

W. Yarberry, *CRAN Recipes*, https://doi.org/10.1007/978-1-4842-6876-6_7

dotall:

```
str_detect("\nX\n", ".X.")
```

```
## [1] FALSE
```

```
str_detect("\nX\n", regex(".X.", dotall = TRUE))
```

```
## [1] TRUE
my.regex <- ".Z."
x <- "aZa"
```

Something on both sides of Z:

```
str_detect(x, regex(my.regex, dotall = TRUE))
```

```
## [1] TRUE
x <- "Za"
str_detect(x, regex(my.regex, dotall = TRUE))
```

```
## [1] FALSE
```

7.2 Anchors

Anchors, like their ship counterpart, provide an attachment point. With an anchor character, a regular expression begins from either the start or end of a string. For example, you may be searching through 10,000 invoices looking for any vendor that starts with "IBM." You would want to see "IBM Parts Division," "IBM Software Services," and so on. You are not interested in "Southern IBMATTER Corporation." Table 7-1 is a summary of anchor characters. Figure 7-1 is a picture you can use to cement your memory on what starts and ends a pattern.

Table 7-1. *Anchor syntax*

Character	What it matches
^	Start of a string.
$	End of a string.
\A	Start of input for a multiline string (multiline = TRUE).
\z	End of input for a multiline string (multiline = TRUE).
\Z	End of the input, but before the final line terminator, if it exists (multiline = TRUE).

Figure 7-1. *Anchors signifying start and end of an applicable pattern*

7.3 Multiline Specification

The character vector x has three line breaks. Only the first instance of the line is extracted, since the other two are on different lines.

```
x <- "Line 1\nLine 2\nLine 3\n"
```

```
str_extract_all(x, "^Line..")[[1]]
```

```
## [1] "Line 1"
```

All three line strings are captured, since multiline is specified:

```
str_extract_all(x, regex("^Line..", multiline = TRUE))[[1]]
```

```
## [1] "Line 1" "Line 2" "Line 3"
```

```
str_extract_all(x, regex("\\ALine..", multiline = TRUE))[[1]]
```

```
## [1] "Line 1"
```

Only one "Line" is extracted, since "\A" indicates extract only the start of a multiline string.

7.4 Word Boundaries

A word boundary can occur in one of three positions:[2]

- Before the first character in the string if the first character is a word character

- After the last character in the string if the last character is a word character

- Between two characters in the string, where one is a word character and the other is not a word character

There are two word boundary notations using "\":

- \b for a word boundary

- \B for not a word boundary

---

[2]"What Is a Word Boundary in RegEx?", Stack Overflow, accessed on March 15, 2020, `https://stackoverflow.com/questions/1324676/what-is-a-word-boundary-in-regex`.

For this code, R requires a double slash. \\b means a whole word anchor (start and stop). Note that only "time" is captured, not "timex":

```
x <- "My timex shows the correct time     "
str_extract(x, "\\btime\\b")

## [1] "time"
str_extract (x, "\\Btimex\\B") #wrong

## [1] NA
string1 <- "Mytimexshows the correct time     "
str_extract(string1, "\\Btimex\\B")

## [1] "timex"
```

Using the same RegEx, a match was found on "time" in string1, because "time" is not surrounded by spaces, a word boundary.

A match is shown but only the first one is output:

```
string2 <- "Mytimexisagenuinetimexforsure!"
str_extract (string2, "\\Btimex\\B")

## [1] "timex"
```

In the first example, "time" is sandwiched between the whole word "My" and "x." In the second example, the pattern to be matched is timex without a word boundary on either side. For the third example, "timex" is sandwiched between two \\B non-word boundaries, and "timex" is the output.

7.5 Whitespace

The most common forms of whitespace you will use with regular expressions are space (␣), tab (\t), newline (\n), and carriage return (\r) (useful in Windows environments). These special characters match each of their respective whitespaces. In addition, a whitespace special character \s will match any of the preceding specific whitespaces and is useful when dealing with raw input text.[3] For example, when you import a text file or csv with many unwanted extra spaces, you can replace many contiguous spaces with a single space.

```
test.vector <- "I pressed the space bar          too many times"
gsub("\\s+"," ",test.vector)
```

```
## [1] "I pressed the space bar too many times"
```

\\s+ will match any space character or repeats of space characters and will replace it with a single space " ".

Built-in terms such as [:space:] can be used to make the code easier to read:

```
some.text <- c("tree","frog   ","House99",
   "Claudia_Carol_Cathy")
i <- grep("[[:space:][:digit:]]+", some.text)
some.text[i]
```

```
## [1] "frog   "   "House99"
```

Frog has a trailing space and House has a trailing digits of 99. One or more spaces or digits were matched with some.text.

7.6 Extended Regular Expressions

Certain named classes of characters are predefined. Their interpretation depends on the POSIX locale. A locale object tries to capture all the defaults that can vary between countries/geographic areas. You set the locale in once, and the details are automatically passed on down to the column parsers. The defaults have been chosen to match R (i.e., US English) as closely as possible.[4]

In R, some classes (groupings) of characters are conveniently predefined.[5] Table 7-2 shows the most common classes.

---

[3]"RegexOne – Learn Regular Expressions – Lesson 9: All This Whitespace," accessed on March 16, 2020, https://regexone.com/lesson/whitespaces.

[4]"Locale Function | R Documentation," accessed on March 26, 2020, www.rdocumentation.org/packages/readr/versions/1.3.1/topics/locale.

[5]"R: Regular Expressions as Used in R," accessed on March 14, 2020, https://stat.ethz.ch/R-manual/R-devel/library/base/html/regex.html.

Table 7-2. *Predefined, extended regular expression terms*

Class	Meaning
[:alnum:]	Alphanumeric characters: [:alpha:] and [:digit:].
[:alpha:]	Alphabetic characters: [:lower:] and [:upper:].
[:blank:]	Blank characters: space and tab and possibly other locale-dependent characters such as non-breaking space.
[:cntrl:]	Control characters. In ASCII, these characters have octal codes 000–037 and 177 (DEL). In another character set, these are the equivalent characters, if any.
[:digit:]	Digits: 0 1 2 3 4 5 6 7 8 9.
[:graph:]	Graphical characters: [:alnum:] and [:punct:].
[:lower:]	Lowercase letters in the current locale.
[:print:]	Printable characters: [:alnum:], [:punct:], and space.
[:punct:]	Punctuation characters: ! " # $ % & ' ( ) * + , . / : ; < = > ? @ [\ ] ^ _ ` {	} ~
[:space:]	Space characters: tab, newline, vertical tab, form feed, carriage return, and space.
[:upper:]	Uppercase letters in the current locale.
[:xdigit:]	Hexadecimal digits: 0 1 2 3 4 5 6 7 8 9 A B C D E F a b c d e f.

7.7 How to Set Locale

As you can see in the following, locale (requires the package readr) sets up parameters specific to a country. locale() resets to R "out-of-the-box" values:

```
locale()

## <locale>
## Numbers:  123,456.78
## Formats:  %AD / %AT
## Timezone: UTC
## Encoding: UTF-8
## <date_names>
## Days:    Sunday (Sun), Monday (Mon), Tuesday (Tue), Wednesday (Wed),
```

```
##          Thursday (Thu), Friday (Fri), Saturday (Sat)
## Months: January (Jan), February (Feb), March (Mar), April (Apr), May
##          (May), June (Jun), July (Jul), August (Aug), September
##          (Sep), October (Oct), November (Nov), December (Dec)
## AM/PM:   AM/PM
locale ("fr")
## <locale
## Numbers:  123,456.78
## Formats:  %AD / %AT
## Timezone: UTC
## Encoding: UTF-8
## <date_names>
## Days:   dimanche (dim.), lundi (lun.), mardi (mar.), mercredi (mer.),
##         jeudi (jeu.), vendredi (ven.), samedi (sam.)
## Months: janvier (janv.), fÃ©vrier (fÃ©vr.), mars (mars), avril (avr.),
mai
##         (mai), juin (juin), juillet (juil.), aoÃ»t (aoÃ»t),
##         septembre (sept.), octobre (oct.), novembre (nov.),
##         dÃ©cembre (dÃ©c.)
## AM/PM:  AM/PM
```

South American locale:

```
locale ("es", decimal_mark = ",")

## <locale>
## Numbers:  123.456,78
## Formats:  %AD / %AT
## Timezone: UTC
## Encoding: UTF-8
## <date_names>
## Days:   domingo (dom.), lunes (lun.), martes (mar.), miÃ©rcoles (miÃ©.),
##         jueves (jue.), viernes (vie.), sÃ¡bado (sÃ¡b.)
## Months: enero (ene.), febrero (feb.), marzo (mar.), abril (abr.), mayo
##         (may.), junio (jun.), julio (jul.), agosto (ago.),
##         septiembre (sept.), octubre (oct.), noviembre (nov.),
##         diciembre (dic.)
## AM/PM:  a. m./p. m.
```

## CHAPTER 8

Elements of Regular Expressions

Literals are simply characters themselves, such as "a" or "boat" or "123." Some characters are "reserved" with special meanings, such as "+." In the case of the plus sign, its special meaning is "additional characters like one just to the left of the + sign." If you want to use any of these reserved characters as a literal in a RegEx, you need to escape them with a backslash. If you want to match 1+1=2, the correct RegEx is 1\\+1=2. Otherwise, the plus sign has a special meaning.[1] Remember that two backslashes are required.

```
string1 <- "This is elementary Watson. 1+1=2"
my.regex <- "1\\+1=2"
my.regex.replacement.value <- "two plus two equals four "
sub(pattern = my.regex,replacement =
  my.regex.replacement.value,x = string1)

## [1] "This is elementary Watson. two plus two equals four "
```

Use a pattern containing only a literal. sub does a replacement when it finds a match with the pattern. Note in this case only the first "a" is replaced:

```
string1  <- "Why do Java Programmers have to wear glasses?
  Because they can't C."
my.regex <- "[a]"
my.regex.replacement.value <- "dog"
```

---

[1] "Regular Expressions Quick Start," accessed on March 23, 2020, www.regular-expressions.info/quickstart.html.

© William Yarberry 2021
W. Yarberry, *CRAN Recipes*, https://doi.org/10.1007/978-1-4842-6876-6_8

```
sub(pattern = my.regex,replacement =
  my.regex.replacement.value,x = string1)
```

```
## [1] "Why do Jdogva Programmers have to wear glasses?  Because they can't C."
```

8.1 Meta-characters

Table 8-1 lists meta-characters and their meanings.

Table 8-1. *Meta-characters and their meanings*

Character	Used to indicate
^	Beginning of string.
$	End of string.
.	Any character except newline.
*	Match zero or more times.
+	Match one or more times.
?	Match zero or one time, or shortest match.
|	Alternative.
( )	Grouping, "storing."
[ ]	Set of characters.
{ }	Repetition modifier.
\	Quote or special.

The character $1 refers to anything inside the first set of *parentheses*.

8.2 Ranges

RegEx ranges are a bit odd. One would think that [0-256] would be the correct RegEx for digits 0–256. Unfortunately, it is not that simple. Sometimes RegEx reminds you of a cousin with bad breath. You tolerate him because he is family but try not to stand too close. Similarly, we must tolerate numeric range RegEx syntax.

8.2.1 Numeric

To check for numeric, use [0-9]. For a specific number of digits (and only that many digits), use ^[0-9]{n}$ where n is the number of digits you need to match. For example, a three-digit number will be matched by ^[0-9]{3}$.

Examples of numeric matches, using grepl. To be true, the string must have only numbers, ranging from one to three digits:

```
x <- c("a","99","89x", "2","333","123", "123456","367")
grepl("^[0-9]{1,3}$",x)
```

```
## [1] FALSE  TRUE FALSE  TRUE  TRUE  TRUE FALSE  TRUE
```

Evaluates to TRUE only if exactly three elements match:

```
grepl("^[0-9]{3}$",x)
```

```
## [1] FALSE FALSE FALSE FALSE  TRUE  TRUE FALSE  TRUE
```

Includes numbers with at least one 0–5 digit:

```
grepl("[0-5]",x)
```

```
## [1] FALSE FALSE FALSE  TRUE  TRUE  TRUE  TRUE  TRUE
```

```
my.regex <- "^([012]?[0-9]?[0-9]|3[0-5][0-9]|36[0-6])$"  #0-366, only
digits
grepl(my.regex,x)
```

```
## [1] FALSE  TRUE FALSE  TRUE  TRUE  TRUE FALSE FALSE
```

Evaluates TRUE only for 0–255 digits:

```
my.regex <- "^([01]?[0-9]?[0-9]|2[0-4][0-9]|25[0-5])$"
grepl(my.regex,x)
```

```
## [1] FALSE   TRUE FALSE   TRUE FALSE   TRUE FALSE FALSE
```

8.2.2 Alpha

grepl, using RegEx, allows text queries, as one would expect.

String must *start with* a lowercase letter to evaluate as TRUE:

```
x <- c("a","99","89x", "2","333","qqq123", "123456z","367")
grepl("^[a-z]",x)
```

```
## [1]   TRUE FALSE FALSE FALSE FALSE   TRUE FALSE FALSE
```

8.3 Case Sensitivity

Example of case sensitivity:

```
stringa <- "OX"
stringb <- "Post Box"
grep(stringa,stringb)
```

```
## integer(0)
```

Now ignore case. The 1 indicates a match of OX with the first (and only) element of stringb:

```
grep(stringa, stringb, ignore.case = TRUE)
```

```
## [1] 1
```

8.4 Repetition

This RegEx pattern matches anything with three or more a's:

```
x <- c("a", "aa", "aaa", "aaaa", "aaaaa")
my.regex <- "a{3}"
grepl(my.regex, x)
```

```
## [1] FALSE FALSE  TRUE  TRUE  TRUE
```

Find repeated words. The RegEx pattern includes both upper- and lowercase letters:

```
x <- c("max", "head head", "HEAD HEAD", "max Max max max")
my.regex <- "\\b([A-Z,a-z]+)\\s+\\1\\b"
grepl(my.regex, x, perl = TRUE)
```

```
## [1] FALSE  TRUE  TRUE  TRUE
```

8.5 Negations: NOT Syntax

The exclamation point indicates negation. In the following example, the entire grepl expression/function is negated.[2] Another way of introducing negation, kindly provided by @IceCreamToucan,[3] is to use "invert = TRUE":

```
string = c("business_metric_one","business_metric_one_dk","business_metric_
one_none","business_metric_two","business_metric_two_dk","business_metric_
two_none")
string[!grepl("business_metric_[[:alpha:]]+_(dk|none)", string)]
```

```
## [1] "business_metric_one" "business_metric_two"
```

[2]"R: RegEx for Containing Pattern with Negation," Stack Overflow, accessed on March 26, 2020, https://stackoverflow.com/questions/46898699/r-regex-for-containing-pattern-with-negation.

[3]"RegEx – Regular Expression Negation in R," Stack Overflow, accessed on March 26, 2020, https://stackoverflow.com/questions/59139892/regular-expression-negation-in-r.

The following task is to find a regular expression which will match only "a rheumatic fever". "invert = TRUE" is a clever way to say "return what grep does not find."

```
my_strings <- c("a non-rheumatic fever",
   "a nonrheumatic fever", "a rheumatic fever",
   "a not rheumatic fever")
grep("no[nt][- ]?rheumatic", my_strings, invert = TRUE,
  value = TRUE)

## [1] "a rheumatic fever"
```

8.6 Grouping

You can use round brackets (parentheses) to bookend part of a regular expression. For example, (\\d+) specifies a contiguous group of numbers. Using the parentheses allows the application of a quantifier to the entire group or to restrict alternation to part of the RegEx.

Regmatches example[4]: Retrieve 1234 and 567 when they are preceded by xy:

```
s <- "xy1234wz98xy567"
r <- "xy(\\d+)"
gsub(r, "\\1", regmatches(s,gregexpr(r,s))[[1]])

## [1] "1234" "567"
```

Use Stringr to extract matched groups from a string:[5]

```
strings <- c(" 219 733 8965", "329-293-8753 ", "banana",
   "595 794 7569","387 287 6718", "apple", "233.398.9187   ",
   "482 952 3315","239 923 8115 and 842 566 4692",
   "Work: 579-499-7527", "$1000","Home: 543.355.3679")
phone <- "([2-9][0-9]{2})[- .]([0-9]{3})[- .]([0-9]{4})"

str_extract (strings, phone)
```

---

[4]"RegEx – Extract Capture Group Matches from Regular Expressions? (Or: Where Is Gregexec?)," Stack Overflow, accessed on March 26, 2020, https://stackoverflow.com/questions/18620571/extract-capture-group-matches-from-regular-expressions-or-where-is-gregexec.

[5]"R: Extract Matched Groups from a String," accessed on March 26, 2020, https://jangorecki.gitlab.io/data.cube/library/stringr/html/str_match.html.

```
##   [1] "219 733 8965" "329-293-8753" NA              "595 794 7569"
##   [5] "387 287 6718" NA             "233.398.9187" "482 952 3315"
##   [9] "239 923 8115" "579-499-7527" NA              "543.355.3679"
```

str_match (strings, phone)

```
##           [,1]            [,2]   [,3]   [,4]
##   [1,] "219 733 8965" "219"  "733"  "8965"
##   [2,] "329-293-8753" "329"  "293"  "8753"
##   [3,] NA             NA     NA     NA
##   [4,] "595 794 7569" "595"  "794"  "7569"
##   [5,] "387 287 6718" "387"  "287"  "6718"
##   [6,] NA             NA     NA     NA
##   [7,] "233.398.9187" "233"  "398"  "9187"
##   [8,] "482 952 3315" "482"  "952"  "3315"
##   [9,] "239 923 8115" "239"  "923"  "8115"
## [10,] "579-499-7527" "579"  "499"  "7527"
## [11,] NA             NA     NA     NA
## [12,] "543.355.3679" "543"  "355"  "3679"
```

Extract/match all:

str_extract_all(strings, phone)

```
## [[1]]
## [1] "219 733 8965"
##
## [[2]]
## [1] "329-293-8753"
##
## [[3]]
## character(0)
##
## [[4]]
## [1] "595 794 7569"
##
## [[5]]
## [1] "387 287 6718"
```

```
##
## [[6]]
## character(0)
##
## [[7]]
## [1] "233.398.9187"
##
## [[8]]
## [1] "482 952 3315"
##
## [[9]]
## [1] "239 923 8115" "842 566 4692"
##
## [[10]]
## [1] "579-499-7527"
##
## [[11]]
## character(0)
##
## [[12]]
## [1] "543.355.3679"
str_match_all(strings, phone)

## [[1]]
##      [,1]           [,2] [,3] [,4]
## [1,] "219 733 8965" "219" "733" "8965"
##
## [[2]]
##      [,1]           [,2] [,3] [,4]
## [1,] "329-293-8753" "329" "293" "8753"
##
## [[3]]
##      [,1] [,2] [,3] [,4]
##
## [[4]]
##      [,1]           [,2] [,3] [,4]
```

```
## [1,] "595 794 7569" "595" "794" "7569"
##
## [[5]]
##       [,1]              [,2]  [,3]  [,4]
## [1,] "387 287 6718" "387" "287" "6718"
##
## [[6]]
##       [,1] [,2] [,3] [,4]
##
## [[7]]
##       [,1]              [,2]  [,3]  [,4]
## [1,] "233.398.9187" "233" "398" "9187"
##
## [[8]]
##       [,1]              [,2]  [,3]  [,4]
## [1,] "482 952 3315" "482" "952" "3315"
##
## [[9]]
##       [,1]              [,2]  [,3]  [,4]
## [1,] "239 923 8115" "239" "923" "8115"
## [2,] "842 566 4692" "842" "566" "4692"
##
## [[10]]
##       [,1]              [,2]  [,3]  [,4]
## [1,] "579-499-7527" "579" "499" "7527"
##
## [[11]]
##       [,1] [,2] [,3] [,4]
##
## [[12]]
##       [,1]              [,2]  [,3]  [,4]
## [1,] "543.355.3679" "543" "355" "3679"
```

Further examples of string matching:

```
x <- c("<a> <b>", "<a> <>", "<a>", "", NA)
str_match (x, "<(.*?)> <(.*?)>")
```

```
##        [,1]        [,2] [,3]
## [1,] "<a> <b>" "a"  "b"
## [2,] "<a> <>"  "a"  ""
## [3,] NA         NA   NA
## [4,] NA         NA   NA
## [5,] NA         NA   NA
str_match_all(x, "<(.*?)>")
```

```
## [[1]]
##        [,1]  [,2]
## [1,] "<a>" "a"
## [2,] "<b>" "b"
##
## [[2]]
##        [,1]  [,2]
## [1,] "<a>" "a"
## [2,] "<>"  ""
##
## [[3]]
##        [,1]  [,2]
## [1,] "<a>" "a"
##
## [[4]]
##        [,1] [,2]
##
## [[5]]
##        [,1] [,2]
## [1,] NA    NA
str_extract(x, "<.*?>")
```

```
## [1] "<a>" "<a>" "<a>" NA     NA
str_extract_all(x, "<.*?>")
```

```
## [[1]]
## [1] "<a>" "<b>"
##
## [[2]]
## [1] "<a>" "<>"
##
## [[3]]
## [1] "<a>"
##
## [[4]]
## character(0)
##
## [[5]]
## [1] NA
```

8.7 Alternation: OR Syntax

The RegEx literature is fond of the Latinate term "alternation." I'm sure there is some deep, mathematical justification for the word. Meanwhile, "OR" is more intuitive for most of us.[6]

8.7.1 The Alternation Operator (l or \l)

Alternatives match one of a choice of regular expressions: if you put the character(s) representing the alternation operator between any two regular expressions a and b, the result matches the union of the strings that a and b match. For example, supposing that `|' is the alternation operator, then `foo|bar|quux' would match any of `foo', `bar', or `quux'.

The alternation OR (as a concept, not literally OR) operates on the largest possible surrounding regular expressions. (Put another way, it has the lowest precedence of any regular expression operator.) Thus, the only way you can delimit its arguments is to use grouping. For example, if `(' and `)' are the open- and close-group operators, `fo(o|b)ar' would match either `fooar' or `fobar'. (`foo|bar' would match `foo' or `bar'.)

The matcher usually tries all combinations of alternatives so as to match the longest possible string. For example, when matching `(fooq|foo)*(qbarquux|bar)'` against `fooqbarquux', it cannot take, say, the first ("depth-first") combination it could match, since then it would be content to match just `fooqbar'.

8.8 Quantifiers

Quantifiers specify the number of characters in the pattern. Applied to the immediately preceding term: ? matches one or zero times; + is for one or more matches; * specifies a match zero or one time. The question mark means that the preceding character(s) are allowed to appear but don't have to. It only attaches to the preceding characters.

Match elements in the string which have zero or one space only:

```
x <- c("aaa","one space","multiple      spaces", " bbbb ")
my.regex <- "^\\S+(?: \\S+)*$"
grepl(my.regex,x)
```

```
## [1]  TRUE  TRUE FALSE FALSE
```

Require one or more spaces:

```
my.regex <- "\\s"
grepl(my.regex, x)
```

```
## [1] FALSE  TRUE  TRUE  TRUE
```

Require at least n but not more than m times:

```
my.regex <- "\\s{2,100}"
grepl(my.regex,x)
```

```
## [1] FALSE FALSE  TRUE FALSE
```

Require two contiguous spaces:

```
my.regex <- "\\s{2,100}"
grepl(my.regex,x)
```

```
## [1] FALSE FALSE  TRUE FALSE
```

8.9 Case Sensitivity

R RegExes default to case sensitive. Placing a (?i) in front of the relevant expression makes it case insensitive.

The grepl default is case sensitive.

```
x <- c("aaa","BBB","ccCC")
my.regex <- "A"
grepl(my.regex,x)
```

```
## [1] FALSE FALSE FALSE
```

Put a (?i) in front of RegEx statements to make them case insensitive:

```
my.regex <- "(?i)A"
grepl(my.regex,x)
```

```
## [1]  TRUE FALSE FALSE
```

8.10 Partial Match

Partial matches are useful if you are looking to match strings which are different but have some common elements:

```
x <- c("miles", "smiles", "milo","smite", "mi lovely vacation")
my.regex <- "mi*l"
grepl(my.regex, x)
```

```
## [1]  TRUE  TRUE  TRUE FALSE FALSE
```

8.11 The Alternation "|" Meta-character

Use the alternation (OR):

```
x <- c("aaa","one space","multiple       spaces", " bbbb ")
my.regex <- "a|p"
grepl(my.regex,x)
```

```
## [1]  TRUE   TRUE   TRUE FALSE
my.regex <- "\\s+"
grepl(my.regex,x)

## [1] FALSE   TRUE   TRUE   TRUE
```

8.12 Look Ahead and Look Behind

Look ahead and look behind, collectively called "lookaround," are zero-length assertions just like the start-end of line and start-end of word anchors. The difference is that lookaround actually matches characters, but then gives up the match, returning only the result: match or no match. That is why they are called "assertions." They do not consume characters in the string, but only assert whether a match is possible or not. Lookaround allows you to create regular expressions that are impossible to create without it or that would get very long-winded without it.[7]

8.12.1 Starter Examples

Look ahead. Find one or more digits, followed by "pencils":

```
x <- c("2 pencils", "3 pencils", "4")
str_extract (x, "\\d+(?= pencils?)")

## [1] "2" "3" NA
```

Look behind. Includes a digit preceded by a dollar sign:

```
y <- c("900", "$500")
str_extract(y, "(?<=\\$)\\d+")

## [1] NA    "500"
```

---

[7]"RegEx Tutorial – Look ahead and Look behind Zero-Length Assertions," accessed on March 16, 2020, www.regular-expressions.info/lookaround.html.

8.12.2 Structure of Look Ahead/Look Behind[8]

- (?=...) – Positive look-ahead assertion. Matches if ... matches at the current input.

- (?!...) – Negative look-ahead assertion. Matches if ... does not match at the current input.

- (?<=...) – Positive look-behind assertion. Matches if ... matches text preceding the current position, with the last character of the match being the character just before the current position. Length must be bounded (i.e., no * or +).

- (?<!...) – Negative look-behind assertion. Matches if ... does not match text preceding the current position. Length must be bounded (i.e., no * or +).

8.12.3 Easy-to-Use and Easy-to-Understand Sets of Characters

These *predefined* classes of characters make R code/regular expressions easier to write. The following is a list of the classes, based on the R manual[9]:

- [:alnum:] – Alphanumeric characters: [:alpha:] and [:digit:].

- [:alpha:] – Alphabetic characters: [:lower:] and [:upper:].

- [:blank:] – Blank characters: space and tab and possibly other locale-dependent characters such as non-breaking space.

- [:cntrl:] – Control characters. In ASCII, these characters have octal codes 000–037 and 177 (DEL). In another character set, these are the equivalent characters, if any.

---

[8]"Regular Expressions," accessed on March 17, 2020, `https://stringr.tidyverse.org/articles/regular-expressions.html`.

[9]"R: Regular Expressions as Used in R," accessed on March 18, 2020, `https://stat.ethz.ch/R-manual/R-devel/library/base/html/regex.html`.

- [:digit:] – Digits: 0 1 2 3 4 5 6 7 8 9.

- [:graph:] – Graphical characters: [:alnum:] and [:punct:].

- [:lower:] – Lowercase letters in the current locale.

- [:print:] – Printable characters: [:alnum:], [:punct:], and space.

- [:punct:] – Punctuation characters: ! " # $ % & ' ( ) * + , - . / : ; < = > ? @ [\ ] ^ _ ` { | } ~.

- [:space:] – Space characters: tab, newline, vertical tab, form feed, carriage return, space, and possibly other locale-dependent characters.

- [:upper:] – Uppercase letters in the current locale.

- [:xdigit:] – Hexadecimal digits: 0 1 2 3 4 5 6 7 8 9 A B C D E F a b c d e f.

CHAPTER 9

The Magnificent Seven

R uses seven regular expression functions for pattern matching and replacement. If you know how to use these functions with appropriate regular expression patterns, then you have a worthy and efficient toolkit for most data science applications.

Let's look at code using each function.

9.1 grep

9.1.1 Pattern Matching and Replacement

grep: a RegEx pattern which matches string vectors. If "value = TRUE" is specified, matching characters are returned. In other words, if the input = "aabbcc" and the pattern is "aa", the output will be "aabbcc". If "value = FALSE", the output will be the index number of the matching vector elements.

grep finds at least one digit, so the string is considered a match with \\d+:

```
text <- "abc2"
grep("\\d+", text, value = TRUE)

[1] "abc2"
```

Here, no digits were found, so the string does not match the pattern of \\d+:

```
text <- "abc"
grep("\\d+", text, value = TRUE)
```

© William Yarberry 2021
W. Yarberry, *CRAN Recipes*, https://doi.org/10.1007/978-1-4842-6876-6_9

```
character(0)
```

grep shows the element number in the vector containing apple:

```
grep("apple", c("crab apple", "Apple jack", "apple sauce", "apple_store"))
```

```
## [1] 1 3 4
```

Answers the question "is string A in string B?" The answer is 1 because "ox" characters were found within "Post Box":

```
stringa <- "ox"
stringb <- "Post Box"
answer <- grep(stringa, stringb)
answer
```

```
## [1] 1
```

Try with an extra "ox" at the end of "Post Box." The answer is 1 because adding an extra occurrence of "ox" makes no difference:

```
stringb <- "Post Boxox"
answer <- grep(stringa, stringb)
answer
```

```
## [1] 1
```

Put "xx" in place of "ox". Integer zero means that "ox" was not found:

```
stringb <- "Post Bxx"
answer <- grep(stringa,stringb)
answer
```

```
## integer(0)
```

Negation method. Check for the *absence* of a string in another string. The answer = 1 means that "zz" is not in "Post Bxx":

```
stringa <- "zz"
answer <- grep(stringa, stringb, invert = TRUE)
answer
```

```
## [1] 1
```

Use case in pattern. The case option defaults to TRUE answer. There is no match because stringa is uppercase:

```
stringa <- "OX"
stringb <- "Post Box"
answer <- grep(stringa,stringb)
answer
```

```
## integer(0)
```

Find "ox" inside various strings:

```
stringa <- "ox"
stringb.as.vector <- c( "Roxan", "bird", "frog", "moxy","MOXY")
answer <- grep(stringa, stringb.as.vector, value = TRUE)
answer
[1] "Roxan" "moxy"
```

9.1.2 Function Structure

The general structure of grep is as follows:

```
grep(pattern, x, ignore.case = FALSE, perl = FALSE, value = FALSE, fixed =
FALSE, useBytes = FALSE, invert = FALSE)
```

- Pattern – The string you are looking for.

- X – Vector you are testing.

- ignore.case – FALSE if pattern matching is case sensitive and otherwise TRUE.

- perl – FALSE tells grep to use R's versions of regular expressions. TRUE means use Perl syntax. Most people ignore this parameter.

- fixed – FALSE. If set to TRUE, all conflicting arguments are overridden. This parameter is not used often.

- useBytes – TRUE if matching is done byte by byte and otherwise if character by character. Not used often.

- invert – TRUE tells grep to return indices or values for elements that do **not** match.

9.2 grepl

grepl: returns TRUE if a string contains the RegEx pattern and FALSE otherwise. The "l" in grepl means logical. grepl expects a vector and returns a TRUE or FALSE for each element. "\\d" means test for a digit and "+" expands to include multiple digits, for example:

```
test <- "the earth has a radius of 3,986 miles"
digits <- grepl("\\d+", test)
digits
```

```
[1] TRUE
```

Look for elements with the characters "on" somewhere in three strings. This is somewhat analogous to Excel's FIND function, except that instead of either a starting number or NA, you get a logical TRUE or FALSE:

```
test.vector <- c("London","Boston","888Knoxville99")
grepl("*on",test.vector)
```

```
[1]  TRUE  TRUE FALSE
```

Look at test.vector to see if any numbers exist. The regular expression includes "d+" to mean one or more digits:

```
grepl("\\d+",test.vector)
```

```
[1] FALSE FALSE  TRUE
```

If you put a number range with a caret "^" inside the brackets, it means NOT what follows. In this case, the pattern means evaluate true/false, based on the start of the string having one or more digits. At this point, you may justifiedly feel that RegEx has had a dose of lunacy added to the mix. The same character means both **start** and **not**, depending on whether it is inside or outside the brackets. Why would the inventors do such a thing? A cosmic joke perhaps.

```
grepl("^([^0-9])",test.vector)
```

```
[1]  TRUE  TRUE FALSE
```

9.2.1 Whitespace

Whitespace, in regular expressions, means a space, tab, carriage return, line feed, or form feed. The symbol for whitespace is "\s."

This expression is true even if Mary is not bracketed by whitespace:

```
x <- "wwwMarywww"
grepl("Mary|Jane|Sue",x)
```

```
[1] TRUE
```

The meta-character \b is an anchor like the caret and the dollar sign. It matches at a position that is called a "word boundary." In the following example, Mary is surrounded by w's, with no boundary. Hence, the pattern matching is false:

```
grepl("\b(Mary|Jane|Sue)\b",x)
```

```
[1] FALSE
```

If we instead use \\S, the absence of a whitespace before any of the names, the expression is TRUE:

```
grepl("\\S(Mary|jane|Sue)",x)
```

```
[1] TRUE
```

Example of date pattern matching:

```
date1 <- "2008-08-08"
pattern1 <- "\\b(\\d{4})-(\\d{2})-(\\d{2})\\b"
```

```
grepl(pattern1, date1)
```

```
[1] TRUE
```

Frustration alert   If you get an error message like this "… unrecognized escape character…," try a double slash rather than a single slash.

Of course, you can enter the pattern directly in quotes within the grepl function:

```
grepl( "\\b(\\d{4})-(\\d{2})-(\\d{2})\\b", date1)
```

```
[1] TRUE
```

9.3 sub

sub: replaces only the first occurrence of a pattern. It returns a character vector of the same length. Elements not matched are returned unchanged. sub means "replace x with y, within the vector v." Only the first element is replaced in sub. See gsub for multiple replacements.

The following code replaces any string starting with letter a plus adjacent letters with a starting z, plus a z at the end of a string of a's. Only the first match in each string element is replaced:

```
sub("(q+)", "z\\1z", c("qbc", "def", "cbq q", "qq"), perl=TRUE)
```

```
[1] "zqzbc"   "def"       "cbzqz q" "zqqz"
```

Note that the second q in "cbq q" was not affected. If you want all matches within a string to be substituted, use gsub.

sub can be used for character replacement. In new.vector1, the original digits are replaced with a single digit 7.

```
vector1 <- c("department 123", "Think ahea", "99 bottles of
   beverage on the wall")
new.vector1 <- sub("[[:digit:]]","7",vector1)
```

```
new.vector1
```

```
[1] "department 723"                    "Think ahea"
[3] "79 bottles of beverage on the wall"
```

Entirely replace any numbers found with the single digit 7 #use + sign:

```
new.vector1 <- sub("[[:digit:]]+","7",vector1)
new.vector1
```

```
[1] "department 7"                    "Think ahea"
[3] "7 bottles of beverage on the wall"
```

9.4 gsub

gsub: works exactly as sub, except that all occurrences of the pattern are replaced, within a string.

In this example, any character equal to 0-9 is set to null for the test:

```
test.vector <- c("ppp - 82", "g - 67.4", "w = 12.544")
scrubbed.var <- gsub("[^0-9]", "", test.vector)
scrubbed.var
```

```
[1] "82"    "674"   "12544"
```

Replace (substitute) an existing vector with null, but keep anything from 0 to 9, plus the period character:

```
test.vector.with.period <- gsub("[^0-9\\.]", "", test.vector)
test.vector.with.period
```

```
[1] "82"    "67.4"   "12.544"
```

Include only lowercase letters in the gsub output:

```
test.vector.lower.case.alpha <- gsub("[^a-z]", "", test.vector)
test.vector.lower.case.alpha
```

```
[1] "ppp" "g"    "w"
```

Remove numbers, replacing any digit with null. \\d means digit:

```
test.vector.no.numbers <- gsub("\\d", "",test.vector)
test.vector.no.numbers
```

```
[1] "ppp - " "g - ."   "w = ."
```

Remove all spaces. \\s is the whitespace indicator:

```
test.vector.no.spaces <- gsub("\\s","",test.vector)
test.vector.no.spaces
```

```
[1] "ppp-82"   "g-67.4"    "w=12.544"
```

9.4.1 More gsub Examples

```
x <- "Lawrence of Arabia"
gsub("Arabia","Chicken scratch Tennessee",x)
```

```
## [1] "Lawrence of Chicken scratch Tennessee"
```

Now attempt a replace with a lowercase a in Arabia. It fails and x remains unchanged:

```
gsub("arabia", "Chicken scratch Tennessee",x)
```

```
## [1] "Lawrence of Arabia"
```

Ignore case and the substitution works again:

```
gsub("arabia","Chicken scratch Tennessee", x,ignore.case=T)
```

```
## [1] "Lawrence of Chicken scratch Tennessee"
```

Replace numbers with asterisks:

```
x <- "HMS Victory is a 104-gun ship of the line of the Royal Navy, launched
in 1765."
y <- gsub("\\d+","***",x)
```

y

```
## [1] "HMS Victory is a ***-gun ship of the line of the Royal Navy,
launched in ***."
```

9.5 regexpr

regexpr: The regexpr function determines where a match is found in a string. The first element of the output shows the starting position of the match in each element. A value of –1 means no match. The second element, "match length," shows the length of the match. In the following first example, the match length is 2, since "ve" is a two-character string. Other patterns will not have such an obvious match length. The third element (attribute "useBytes") has a value TRUE, meaning matching was done byte by byte rather than character by character:[1]

```
x <- c("version 1.2", "vvv24", "engine is a v8", "bigfoot", "have you been
there?")
regexpr("ve", x)
```

```
## [1]  1 -1 -1 -1  3
## attr(,"match.length")
## [1]  2 -1 -1 -1  2
## attr(,"index.type")
## [1] "chars"
## attr(,"useBytes")
## [1] TRUE
```

```
x <- "Alexander the Great was born in 356 BCE"
y <- regexpr("\\d+",x)
y
```

```
## [1] 33
## attr(,"match.length")
## [1] 3
## attr(,"index.type")
```

---

[1] "Regular Expressions with grep, Regexp and sub in the R Language," accessed on February 5, 2020, www.regular-expressions.info/rlanguage.html.

```
## [1] "chars"
## attr(,"useBytes")
## [1] TRUE
```

9.6 gregexpr

gregexpr: The gregexpr function is used to identify where a pattern is within a character vector, where each element is searched separately. The returned object is a list. In contrast, the regexpr function returns a vector, for example:

```
v1  <- " Do rats prefer pie 3.141 What does the golden rat prefer 1.6180 "
v2  <- gregexpr("\\d+",v1)
v2
```

```
## [[1]]
## [1] 21 23 59 61
## attr(,"match.length")
## [1] 1 3 1 4
## attr(,"index.type")
## [1] "chars"
## attr(,"useBytes")
## [1] TRUE
```

Here's how to interpret this output:

- Length result – In this example, the first number 3 has a length of 1.

- The second number (after the period) has a length of 3.

- The fourth number, 1, has a length of 1.

- The fifth number, 6180, has a length of 4.

- Index result – Separate (noncontiguous) numbers start in positions 21, 23, 61, and 63.

9.7 regexec

regexec: These functions perform pattern matching, logical tests, identification of string character locations, vector index numbers, replacements, inversion (show what does not match), partial matching, case and case-insensitive matching, and I'm sure a few others buried deep in the technical manuals. strsplit and a few others use regular expressions, but I thought the magnificent seven sounded better than the magnificent eight.

RegEx returns a list of the same length as the text of each element which is either –1 if there is no match or a sequence of integers with the starting positions of the match and all substrings corresponding to parenthesized subexpressions of pattern. It also includes the attribute "match.length", a vector giving the lengths of the matches (or –1 for no match). The interpretation of positions and length and the attributes follows regexpr.[2]

Example[3] separating a URL into its components:

```
x <- "http://stat.umn.edu:80/xyz"
m <- regexec("^(([^:]+)://)?([^:/]+)(:([0-9]+))?(/.*)", x)
m

## [[1]]
## [1]  1  1  1  8 20 21 23
## attr(,"match.length")
## [1] 26  7  4 12  3  2  4
## attr(,"index.type")
## [1] "chars"
## attr(,"useBytes")
## [1] TRUE

regmatches(x, m)

## [[1]]
## [1] "http://stat.umn.edu:80/xyz" "http://"
## [3] "http"                       "stat.umn.edu"
```

---

[2]"grep Function | R Documentation," accessed on March 18, 2020, www.rdocumentation.org/packages/base/versions/3.6.2/topics/grep.

[3]"R: Pattern Matching and Replacement," accessed on March 18, 2020, https://stat.ethz.ch/R-manual/R-devel/library/base/html/grep.html.

CHAPTER 10

Regular Expressions in Stringr

You can use the following prebuilt classes in the package Stringr[1]:

- [:punct:] – Punctuation
- [:alpha:] – Letters
- [:lower:] – Lowercase letters
- [:upper:] – Uppercase letters
- [:digit:] – Digits
- [:xdigit:] – Hexadecimal digits
- [:alnum:] – Letters and numbers
- [:cntrl:] – Control characters
- [:graph:] – Letters, numbers, and punctuation
- [:print:] – Letters, numbers, punctuation, and whitespace
- [:space:] – Space characters (basically equivalent to \s)
- [:blank:] – Space and tab

---

[1]Simple, Consistent Wrappers for Common String Operations. A consistent, simple, and easy-to-use set of wrappers around the fantastic "stringi" package. All function and argument names (and positions) are consistent, all functions deal with "NAs" and zero-length vectors in the same way, and the output from one function is easy to feed into the input of another. Source: "Stringr Package | R Documentation," accessed on March 26, 2020, www.rdocumentation.org/packages/stringr/versions/1.4.0.

W. Yarberry, *CRAN Recipes*, https://doi.org/10.1007/978-1-4842-6876-6_10

Examples:

```
str_extract_all("The Howdy Doody Show 123", "[:alpha:]")
```

```
[[1]]
 [1] "T" "h" "e" "H" "o" "w" "d" "y" "D" "o" "o" "d" "y" "S" "h" "o" "w"
```

```
str_extract_all("The Howdy Doody Show 123", "[:digit:]")
```

```
[[1]]
[1] "1" "2" "3"
```

In both cases, the output is a list. Also note the prebuilt name is "digit," not "digits."

CHAPTER 11

Unicode

11.1 The World of ASCII, UTF-8, Latin-1, and All the Rest

ASCII coding is familiar to most of us. Unfortunately, ASCII is the exception—a simple, plain, "white bread" version of the character world. Latin-1 is less commonly used. Other representations of characters depend on the standard selected, which in turn may depend on the locale. As an example, consider the trademark sign. In MS Word and other MS Office applications, Ctrl-Alt-T or Alt8482 will create the trademark sign: ™. In Unicode, it is \u2122 or outside R is U+2122. In HTML Hex, use ™. Once you determine the coding scheme, finding the values is straightforward, since they are well documented on the Internet. To find information on various Unicodes, see the official website of the Unicode Consortium, `www.Unicode.org`. Another helpful resource is `www.utf8-chartable.de/`.

Unicode is a large topic with entire books devoted to its intricacies. Most R programmers will not need to learn Unicode in depth. Typically, if you are using relatively recent, day-to-day type files and databases, you will not need to research Unicode options. Data created decades ago is likely to have unusual characters, requiring you to jump into a rabbit hole of coding analysis. If so, good luck.

Identify text with trademark (UTF-8).

One trick to get various codes into RStudio is to type the code in another software package and paste. For example, to get a trademark sign into RStudio, I typed control-alt-t in MS Word and copied it into the following line:

© William Yarberry 2021
W. Yarberry, *CRAN Recipes*, https://doi.org/10.1007/978-1-4842-6876-6_11

```
x <- "™ this string starts with a trademark sign"
my.regex <- "\u{2122}"
grepl(my.regex, x)
```

```
## [1] TRUE
```

```
x <- "this string has no trademark sign"
grepl(my.regex, x)
```

```
## [1] FALSE
x <- "This string has a colon in it :  "
my.regex <- "\u{003A}"
grepl(my.regex, x)
```

```
## [1] TRUE
```

To see what a particular Unicode looks like in RStudio, just set a variable equal to it and display

```
x <- "\u00b5"
x
```

```
## [1] "µ"
```

CHAPTER 12

Tools for Development and Resources

12.1 Utilities

12.1.1 regex101.com

The free and easy-to-use website `www.regex101.com` serves as a goto sandbox when you are trying different approaches and need immediate feedback. See Figures 12-1 and 12-2. It is not specifically tuned to R, but the great majority of R RegEx configurations will be properly evaluated. To the right of the screen, explanatory comments are provided. Although it has far fewer features than RegexBuddy (discussed later), for many people, it will be sufficient. Since doing is closely tied to learning, spending a few hours with this tool on the front end will accelerate your learning curve.

© William Yarberry 2021

W. Yarberry, *CRAN Recipes*, https://doi.org/10.1007/978-1-4842-6876-6_12

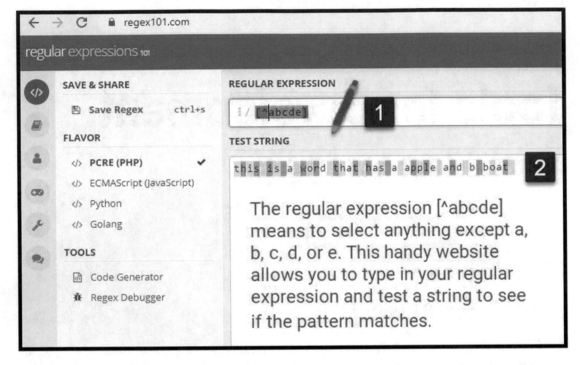

Figure 12-1. *Simple and handy website for validation of any regular expression against a string*

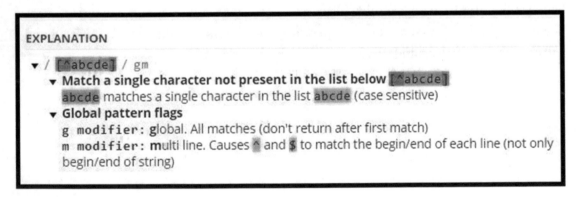

Figure 12-2. *Helpful explanations provided by regex101.com*

12.1.2 RegExBuddy

RegExBuddy is the only RegEx helper I know of that supports and tracks a large number of languages,[1] down to the version level. As of March 2020, its website includes 262 languages/versions. Without question, it is the Cadillac of RegEx utilities. The following are the R versions supported:

- R 2.14.0–2.14.1

- R 2.14.2

- R 2.15.0–3.0.2

- R 3.0.3–3.1.2

- R 3.1.3–3.4.4

- R 3.5.0–3.6.1

- R 4.0.0

The website www.regexbuddy.com/benefits.html lists the following benefits (author's note: I have no financial relationship with RegExBuddy):

- Introduction to regular expressions.

- Learn the regular expression syntax.

- Create regular expressions.

- Understand regular expressions created by others.

- Compare regular expression flavors.

- Convert between regular expression flavors.

- Test regular expressions.

- Debug regular expressions.

- Generate source code snippets.

---

[1]"Compare How Different Applications Interpret a RegEx," accessed on March 24, 2020, www.regexbuddy.com/compare.html.

- Search through files and folders (grep).

- Integrate with other software.

- Runs on Linux using WINE.

Many data scientists, developers, and analysts use both R and Python. There are some differences in RegEx syntax, but the bulk of what you have learned in one language will match the other.

12.1.3 Other Utilities

The following are other websites for RegEx testing. Some include predefined patterns for common data formats, such as Social Security numbers, IDs for various nationalities, and so on. Some are free, but the more full-featured ones charge either a one-time or monthly fee. Before purchase, make sure R is fully supported.

- `http://RegexPal.com`

- `www.nregex.com`

- `www.rubular.com`

- `www.myregexp.com`

- `www.ultrapico.com/Expresso.htm`

- `http://sourceforge.net/projects/regulator`

- `www.debuggex.com/`

12.1.4 Text Editors with Built-In RegEx Capability

RegEx's use across so much software partially offsets its demanding learning curve. For example, a full-featured text editor, Notepad++, permits the use of RegEx for search, replacement, and other functions.

Table 12-1 lists Notepad++-supported programming languages.[2]

---

[2]"Programming Languages | Notepad++ User Manual," accessed on March 24, 2020, `https://npp-user-manual.org/docs/programing-languages/`.

Table 12-1. *Languages supported by Notepad++*

ActionScript	Ada	ASN.1	ASP	Assembly
AutoIt	AviSynth scripts	BaanC	batch files	Blitz Basic
C	C#	C++	Caml	CMake
Cobol	CoffeeScript	Csound	CSS	D
Diff	Erlang	escript	Forth	Fortran
FreeBASIC	Gui4Cli	Haskell	HTML	INI files
Intel HEX	Inno Setup scripts	Java	JavaScript	JSON
JSP	KiXtart	LaTeX	LISP	Lua
Makefile	Matlab	MMIX	Nimrod	nnCron
NSIS scripts	Objective-C	OScript	Pascal	Perl
PHP	PostScript	PowerShell	PureBasic	Python
R	Rebol	Registry script (.reg)	Resource file	Ruby
Rust	Scheme	Shell script	Smalltalk	SPICE
SQL	Swift	S-Record	Tcl	Tektronix HEX
TeX	txt2tags	Visual Basic	Visual Prolog	VHDL
Verilog	XML	YAML		

Figure 12-3 is a screenshot showing RegEx being used in the editor.

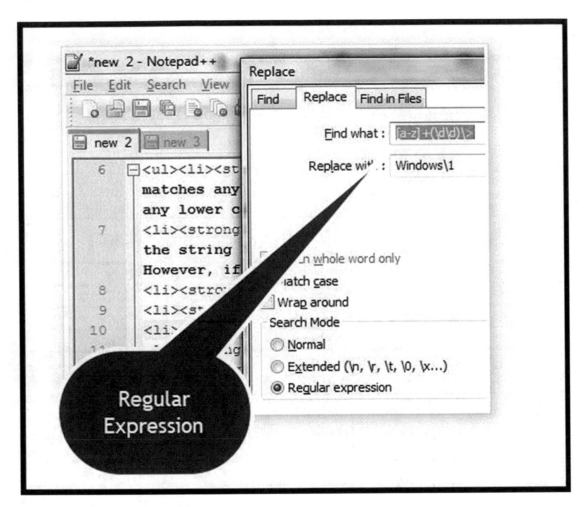

Figure 12-3. *RegEx used inside the text editor Notepad++*

12.1.4.1 Use of RegEx Inside Notepad++

The following notes from Notepad++'s website provide simplified instructions on using RegEx within the application.[3]

---

[3]Haider M. al-Khateeb, "Understanding RegEx with Notepad++ | Dr. Haider M. al-Khateeb," accessed on March 24, 2020, http://blog.hakzone.info/posts-and-articles/editors/ understanding-regex-with-notepad/comment-page-1/#comments.

Searching a string using the "Find" or "Find and Replace" function in text editors highlights the relevant match (e.g., searching "le" highlights it inside words such as "apple," "please," etc.). However, some advanced editors such as Notepad++ (I mention Notepad++ in my examples since it's my favorite so far!) support the use of RegEx, which recently saved me hours of manually replacing strings and numeric values in files containing HTML and JavaScript codes.

RegEx characters can be used to create advanced matching criteria. The following list introduces some of them with practical examples. Before starting, make sure that you change the Search Mode from Normal to Regular expression in your Find or Find and Replace dialog box.

[ ] – The square brackets can be used to match ONE of multiple characters. For instance, [abc] matches any of the characters a, b, or c. Hence, b[eo]n will match words like ben and bon, but not been or beon. Ranges can also be used. [a-z] is any lowercase character and so on.

^ – The caret can be used inside the square brackets to exclude characters from the match. For instance, hell[^o] means the string "hell" will be ignored if followed by the letter "o." Another example is [^A-Za-z] which will exclude all alphabetic characters. However, if not placed inside a set, ^ can be used to match the start of a line.

$ – This matches the end of a line.

. – The period or dot matches any character.

\d – Matches any single digit.

\w – Matches any single alphanumeric characters or underscore.

\s – Matches whitespaces including tabs and line breaks.

* – The asterisk or star sign matches zero or more times. For example, Ba*m matches Bm, Bam, Baam, and so on.

+ – The plus sign matches one or more times. For example, lo+l matches lol, lool, loool, and so on.

\< – Matches the start of a word. For example, \< directly followed by "sh" matches "she" but does not match "wish."

\> – Matches the end of a word. For example, sh\> matches "wish" and does not match "she."

( ) – The round brackets can be used in the Find and Replace function to tag a match. Tagged matches can then be used in replace with \1, \2, and so on.

For example, if you write 123xxxRRR in the search and 123\1HHH in the "Replace with" field, the result will be 123xxxHHH.

\ – The backslash can be used to escape RegEx characters. For example, to match 1+1=2, the correct RegEx is 1\+1=2. Otherwise, the plus sign will have a special meaning.

Further, the following two examples should be giving you a better idea of how to use RegEx in your editor:

Find: Win([0-9]+) Replace with: Windows\1

will search for strings like Win2000 and Win2003 and change them to Windows2000, Windows2003....

Find: [a-z]+(\d\d)\> Replace with: Windows\1

will search for all alphanumerics followed by two digits only at the end such as Win98 and Win07 and change them to Windows98, Windows07....

12.1.5 Regular Expression Capability in Google Sheets

Many utilities and packages throw in regular expression capability as an add-on. The following is an example email formatting from madebyspeak.com[4]:

Here at Speak, we use regular expressions often when dealing with large spreadsheets, database exports, content imports—you name it. If it's longer than a 140-character tweet, someone on our Technology team is going to be reaching for a regular expression from the tool belt to help expedite the change.

Let's start small. You're working on a spreadsheet that looks like Figure 12-4.

	A	B
1	Users	Address
2	james.dean	Sherman Oaks, California
3	bob.saget	9150 Wilshire Blvd. Suite 350, Beverly Hills, CA 90212 USA
4	jacob.savage	8337 Cordova Rd #102, Cordova, TN 38016
5	justin.bieber	211 East 43rd Street Suite 1501 New York, NY 10017 USA
6		

Figure 12-4. *Google Sheets example spreadsheet*

[4]"Become a Master Text Wrangler with Regular Expressions," accessed on March 25, 2020, www.madebyspeak.com/blog/posts/become-a-master-text-wrangler-with-regular-expressions.

The Users column is in the format first.last, but we actually want those to be email addresses. The first part of the email address can contain what's already there, but we want to add @madebyspeak.com to the end of each User so that it forms a valid email address. If this spreadsheet had over one thousand rows, making these updates manually would take a very long time. We wouldn't be able to use a traditional find and replace operation because each cell contains something different. However, because each cell is in the same format (first.last), we can find the last character in the cell and append it with @madebyspeak. com. In Google Sheets, in the Find and Replace tool, there is an option to enable Regular Expressions, as shown in Figure 12-5. To update all the users to email addresses, we would:

- Select the cells we want to update

- Find (.*)$

- Replace with $1@madebyspeak.com

- "Search using regular expressions" and "Match entire cell contents"

- Replace All

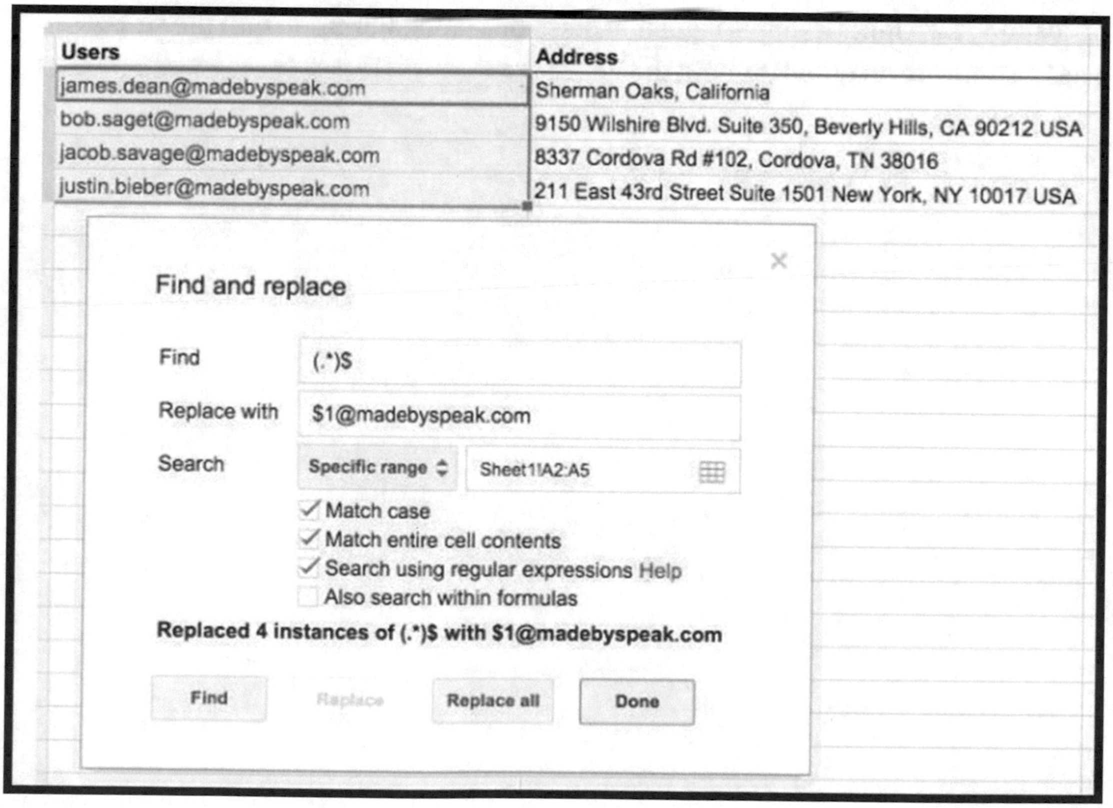

Figure 12-5. *Find and replace*

.* matches any character (.) zero or more times (*). We could have also matched for .+, which would find any character (.) one or more times (+). The ( ) creates a group from this match that we can use later in our replace expression. In the "Replace with" expression, we start with $1, which is a reference to the group created by ( ) in the "Find" expression. We want to replace the whole cell with the content of the original "Find" result plus @madebyspeak.com, so the group $1 that matched (.*) (all characters) will give us the original content, and then we just have to populate what we want to come after the original content—resulting in $1@madebyspeak.com.

12.1.6 RegEx in RStudio

RStudio includes a facility to use regular expressions for find and/or replace text within scripts. This feature is solely for work in your source code. For example, you could use RegEx to change setwd() from "c:\\alpha1" to "f:\\beta1" within your script. This might be overkill for simple substitutions, but if you have more complex changes, RegEx may be a good approach.

Figure 12-6 shows a simple regular expression "[abc]{3}[0-9]" to find the string "aaa1." There are no quotes around the RegEx.

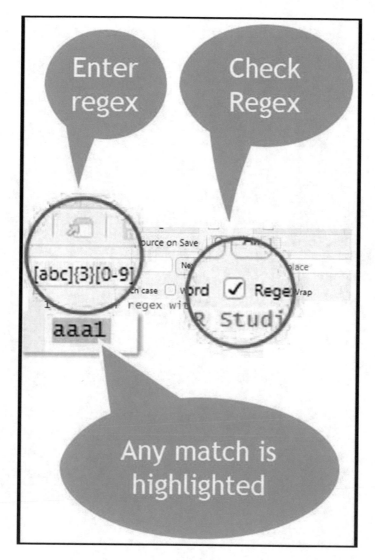

Figure 12-6. *Use of RegEx for search/replacement within RStudio*

## CHAPTER 13

RegEx Summary

Pain, then gain: that's the typical result of learning regular expressions. R includes many functions which duplicate RegEx's capability for specific actions. However, the scope of RegEx pattern matching exceeds traditional R logic. Even knowing just a few RegEx examples will speed your code development and possibly reduce execution time of your script.

My intent is to get you started with relatively simple examples ***which work in R without modification***. With so many flavors of RegEx, finding code that will work in R challenges your Google search skills. Many of the websites do not mention the language when showing examples.

One of my hobbies is studying history, so I'll use an analogy from ancient Rome. At the empire's height, Roman legions routinely conquered other armies, even when outnumbered. Besides discipline and training, they had another advantage over their enemies—they trained on multiple weapons, rather than just one, such as use of the sword. This versatility often saved the day for them. Similarly, having RegEx in your back pocket will serve as a second tool to help you meet your objectives. Besides, it is one of those yes/no credentials which distinguish top- from middle-tier professionals.

© William Yarberry 2021
W. Yarberry, *CRAN Recipes*, https://doi.org/10.1007/978-1-4842-6876-6_13

CHAPTER 14

Recipes for Common R Tasks

Load the following packages to execute the code in subsequent sections:

```
library(tidyverse)
library(readr)
library(datasets)
```

The following packages may not be on your machine unless you have installed them earlier. The following if logic inspects your R libraries to determine whether the package has been installed. If not, it installs them immediately. This eliminates unnecessary reinstalls. This logic always loads the designated package into memory. Some of the examples in this chapter require additional packages, as shown as follows. The syntax shown checks to see if the package is installed on your machine. If it is, the package is simply loaded, as you would do in a library command. If the package is not installed, it installs it and then loads it. If you use R enough, you'll eventually forget whether some of the less frequently used packages have been installed.

```
if (!require("sjPlot")) install.packages("sjPlot")
if (!require("ggcharts")) install.packages("ggcharts")
if (!require("Hmisc")) install.packages("Hmisc")
if (!require("summarytools")) install.packages("summarytools")
if (!require("skimr")) install.packages("skimr")
```

W. Yarberry, *CRAN Recipes*, https://doi.org/10.1007/978-1-4842-6876-6_14

14.1 Input-Output

14.1.1 Console Input

```
my.city.name <- readline(prompt="What is your city name ")
#at this point I entered "knoxville" in the console
my.city.name

##[1] "knoxville"
```

14.1.2 Read and Write CSV Files

Set your working directory, write out the built-in dataset mpg, and then read it back in to demonstrate the read_csv function.

```
setwd("c:\\temp") #set working directory. Your setting will vary
```

Use the mpg built-in dataset:

```
write_csv(mpg, "my.mpg.csv")
y <- read_csv("my.mpg.csv")
head(y,2)

## # A tibble: 2 x 11
##    manufacturer model displ year   cyl trans        drv   cty   hwy fl    class
##    <chr>        <chr> <dbl> <dbl> <dbl> <chr>        <chr> <dbl> <dbl> <chr> <chr>
## 1 audi          a4      1.8  1999     4 auto(l5)     f        18    29 p     compa~
## 2 audi          a4      1.8  1999     4 manual(m5)   f        21    29 p     compa~
```

14.1.3 Windows: Copy a File

If R were a person, we would call her a polymath. It does all the usual analytics and data science stuff but also the mundane tasks required for any programming language. Here, it does a standard Windows file copy of jpegs. First, identify the folder containing the file to be copied. Next, identify the target folder. Then create a list of files having the pattern of ending in "jpeg." Remember from the RegEx chapters that $ is an ending anchor. TRUE means the copy was successful.

Note   This script will not work on Mac. Also, both the starting and ending folder
name need to exist in advance of running the code.

```
starting.folder <- "H:/t"
ending.folder <- "H:/t2"
files.to.be.backed.up <- list.files(starting.folder, ".jpg$")
file.copy(files.to.be.backed.up, ending.folder)
```

```
##[1] TRUE
```

CHAPTER 15

Data Structures

15.1 Built-In Datasets

MASS is a commonly used source of datasets for practicing R code. Figure 15-1 shows a partial list of datasets available. Although it is tempting when you have a new project to jump in using your actual data, consider working through your code first with a toy, built-in dataset so that you are varying only one thing at a time—get the code logic down first, and then work with your own data.

```
library(MASS)
data() #shows base datasets available.
```

```
Data sets in package 'datasets':

AirPassengers           Monthly Airline Passenger Numbers 1949-1960
BJsales                 Sales Data with Leading Indicator
BJsales.lead (BJsales)  Sales Data with Leading Indicator
BOD                     Biochemical Oxygen Demand
CO2                     Carbon Dioxide Uptake in Grass Plants
ChickWeight             Weight versus age of chicks on different diets
DNase                   Elisa assay of DNase
EuStockMarkets          Daily Closing Prices of Major European Stock Indices,
                        1991-1998
Formaldehyde            Determination of Formaldehyde
HairEyeColor            Hair and Eye Color of Statistics Students
Harman23.cor            Harman Example 2.3
Harman74.cor            Harman Example 7.4
Indometh                Pharmacokinetics of Indomethacin
InsectSprays            Effectiveness of Insect Sprays
JohnsonJohnson          Quarterly Earnings per Johnson & Johnson Share
LakeHuron               Level of Lake Huron 1875-1972
LifeCycleSavings        Intercountry Life-Cycle Savings Data
Loblolly                Growth of Loblolly pine trees
Nile                    Flow of the River Nile
Orange                  Growth of Orange Trees
OrchardSprays           Potency of Orchard Sprays
PlantGrowth             Results from an Experiment on Plant Growth
Puromycin               Reaction Velocity of an Enzymatic Reaction
Seatbelts               Road Casualties in Great Britain 1969-84
Theoph                  Pharmacokinetics of Theophylline
```

Figure 15-1. Built-in datasets from package "datasets"

There are all sorts of datasets available. For example, the simple Nile river dataset has a time series format:

```
data(Nile)
head(Nile)
```

```
## [1] 1120 1160   963 1210 1160 1160
```

```
class(Nile)
```

```
## [1] "ts"
```

```
x <- as.data.frame(Nile)
class(x)
```

```
## [1] "data.frame"
```

```
head(x) #demonstrate conversion to dataframe
```

```
##       x
## 1 1120
## 2 1160
## 3   963
## 4 1210
## 5 1160
## 6 1160
```

CHAPTER 16

Visualization

Most of these examples use the R workhorse, ggplot2. The ggplot2 package has had many spin-offs and is a gold mine of logically structured visualizations. The following sections are a smorgasbord of visualizations from ggplot2 and other packages. ggplot2 gets loaded automatically with the library(tidyverse) command.

Although it is good to learn the details of ggplot and other visualization packages, keep in mind that there is a point of diminishing returns when tweaking slight variations in text, annotations, shapes, and so on. It is not "cheating" to do your final visualizations by using a graphics package such as TechSmith's Snagit, Adobe illustrator, Adobe Photoshop, or even Paint to get to your best possible graphic. When you are presenting your results to decision makers, it is unlikely they will say, "Fess up, Sally. I know you didn't do that last annotation in ggplot...."

16.1 Histogram

See Figure 16-1 for an example histogram.

```
ggplot(data=mtcars,aes(mpg))+geom_histogram(aes(y =..density..),
  fill="red4")+geom_density()

## `stat_bin()` using `bins = 30`. Pick better value with `binwidth`.
```

© William Yarberry 2021
W. Yarberry, *CRAN Recipes*, https://doi.org/10.1007/978-1-4842-6876-6_16

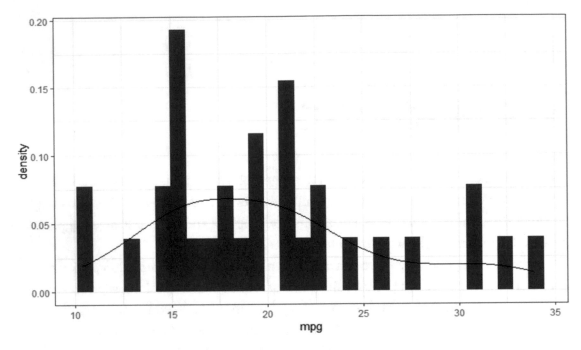

Figure 16-1. *Histogram from ggplot2*

16.2 Chart Variations

Thomas Neitmann's ggcharts include many easy-to-use variations. I adapted the built-in R dataset, txhousing, to supply data for the bar charts. DPLYR summarises sales by city and year. Faceting allows for multiple graphs to be shown together.

16.2.1 Horizontal Bar Charts (Faceted)

See Figure 16-2 for an example faceted horizontal bar chart.

```
library(ggcharts)
df <- txhousing %>%
  group_by(city,year) %>%
  summarise(sales_tot = sum(sales))
```

```
## `summarise()` regrouping output by 'city' (override with `.groups`
argument)

df %>% filter(year %in% c(2012,2013,2014,2015)) %>%
  bar_chart(x = city, y = sales_tot, facet = year, top_n = 10)
```

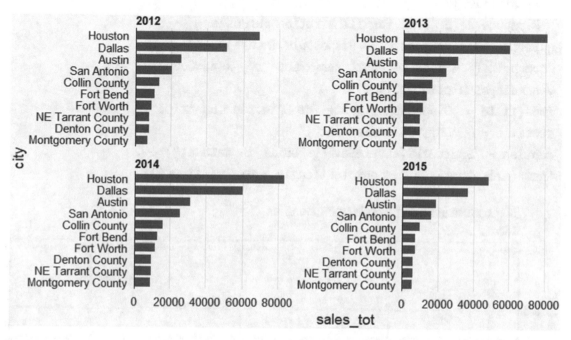

Figure 16-2. *ggcharts: horizontal bar graph. Faceting (4 in 1) is automatic*

16.2.2 Lollipop

Some graphical approaches have an artistic and informal style. Would a lollipop chart appear in the journal *Nature*? Probably not. Would it appear in a magazine or in the annual report of an au courant, hip design company? Possibly. With the growth of R comes the growth of sheer innovation. R's graphics give you the choice of flat-out dull, conservative, middle of the road, creative, or other worldly. Figure 16-3 illustrates disposable income using a lollipop-type ggplot. The dataframe LifeCycleSavings comes from the datasets package.

```
df <- LifeCycleSavings

df <- cbind(Country=rownames(df), df)
  #row names converted to a column
df$disposable.income <- df$dpi
df <- df[1:10,]
  # reduce no of rows for illustration purposes
ggplot(df, aes(x=Country, y=disposable.income)) +
 geom_point() + geom_segment( aes(x=Country, xend=Country, y=0,
 yend=disposable.income)) +
 labs(title = "Disposable Income for Selected Countries",
 subtitle = "Country vrs disp inc",
 caption = "Source:LifeCycleSavings built in dataset") +
 theme(axis.text.x = element_text(angle = 65, vjust=0.6))
```

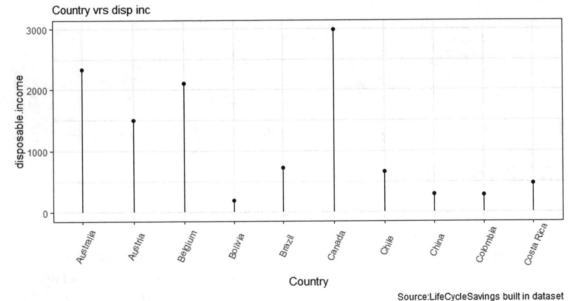

Figure 16-3. *Lollipop chart from ggplot2*

16.2.3 Step Chart

Figure 16-4 shows a step chart of postage price vs. year. The code is from flowingdata. com, a handy, visualization-oriented website run by Nathan Yau[1]:

```
postage = read.csv("http://datasets.flowingdata.com/us-postage.csv", sep =
",", header=T)
head(postage,3); plot(postage$Year, postage$Price, type = "s")

##    Year Price
## 1 1991  0.29
## 2 1995  0.32
## 3 1999  0.33
```

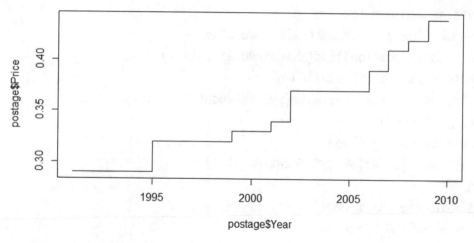

Figure 16-4. *Step chart*

---

[1]Nathan Yau, "FlowingData," FlowingData, accessed on January 31, 2021, https://flowingdata.com/.

16.2.4 Diverging Bars

Figure 16-5 illustrates a diverging bar plot. This ggplot2 powered graphic[2] is great for showing where two items (could be people, business units, treatments, etc.) can be compared on the basis of some numeric value. In the following example, French-speaking provinces in Switzerland are compared based on education levels. The data is from 1888, using the built-in dataset "swiss."

```
data("swiss")
theme_set(theme_bw())
# Data Preparation
data("swiss")  # load data (R built-in dataset)
swiss$`Provence` <- rownames(swiss)
# create new column for Provences
swiss$Education_z <- round((swiss$Education -
  mean(swiss$Education))/sd(swiss$Education), 2)
# compute normalized Education
swiss$Education_type <- ifelse(swiss$Education_z < 0,
 "below", "above")
# above / below avg flag
swiss <- swiss[order(swiss$Education_z), ]
# sort
swiss$`Provence` <- factor(swiss$`Provence`,
  levels = swiss$`Provence`)
# convert to factor to retain sorted order in plot.
# Diverging Barcharts
ggplot(swiss, aes(x=`Provence`, y=Education_z, label="Relative Score in
Education")) +
  geom_bar(stat='identity', aes(fill=Education_type), width=.5)  + scale_
fill_manual(name="Education Level in Years",
  labels = c("Above Average", "Below Average"),
  values = c("above"="#00ba38", "below"="#f8766d")) +
  theme(text = element_text(size=08)) +
```

---

[2]"Divergent Bars in Ggplot2," Once Upon Data, accessed on January 25, 2021, https://onceupondata.com/2019/01/25/ggplot2-divergent-bars/.

```
labs(subtitle="Education Levels by Provence",
    title= "Diverging Bars") +
coord_flip()
```

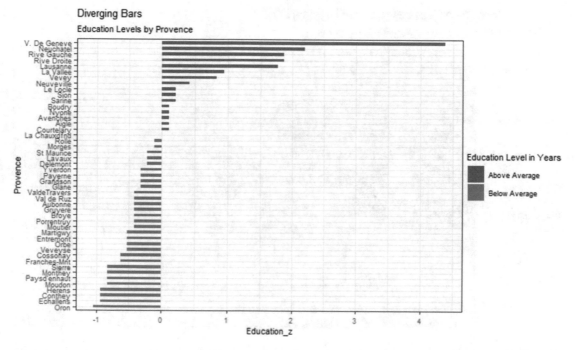

Figure 16-5. *Diverging bars showing relative variance from zero (normalized)*

16.2.5 Colorful Display of Categorical/Character Frequencies

Figure 16-6 shows a quick view of the categorical/character values in a data.frame, using the inspectdf package.[3] The size of the block is proportional to the number of occurrences of the categorical value.

```
if (!require("inspectdf")) install.packages("inspectdf")
library(inspectdf)
```

---

[3]https://cran.r-project.org/web/packages/inspectdf/inspectdf.pdf. Accessed on January 25, 2021.

```
mpg %>%
  inspectdf::inspect_cat() %>%
  inspectdf::show_plot(text_labels = TRUE)
```

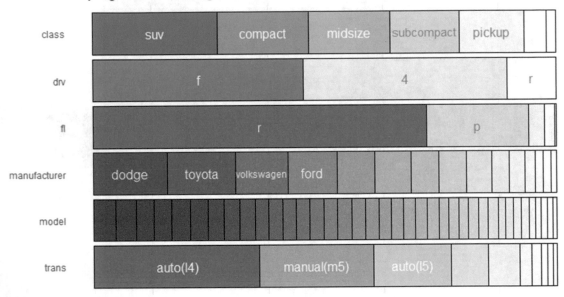

Figure 16-6. *Categorical level frequencies in built-in dataset MPG*

16.2.6 Donut Chart

A donut chart in a scientific paper or a technical financial analysis would be considered a bit amateurish. However, there are occasions when it could be used for special effect, where exact numbers are not critical. The code is simple. Figure 16-7 shows a simple chart without color. Figure 16-8 is an improved version with color added.

```
if (!require("ggpubr")) install.packages("ggpubr")
df <- data.frame(supplies = c("Pencils", "Paper", "Staples", "Elmer's
Glue"), value = c(50, 200, 122, 5))
ggdonutchart(df, "value", label = "supplies")
```

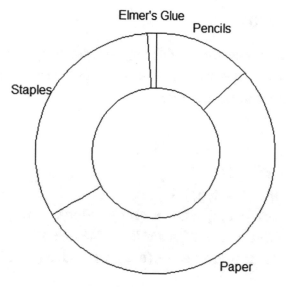

Figure 16-7. *Donut chart with no color*

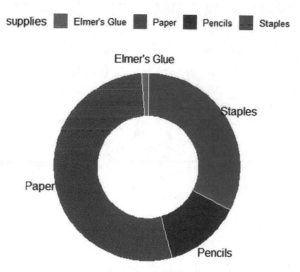

Figure 16-8. *Donut chart with color added*

16.2.7 Bubble Plot

Figure 16-9 uses ggplot to show the relationship between mpg and weight for the built-in dataset mtcars.[4] qsec, the quarter mile time, changes the size of the bubble, and the number of cylinders determines the color:

```
data("mtcars")
df <- mtcars
df$cyl <- as.factor(df$cyl)
ggplot(df, aes(x = wt, y = mpg)) +
  geom_point(aes(color = cyl, size = qsec), alpha = 0.5) +
  scale_color_manual(values = c("#00bb06", "#E7B800",
  "#FC4E07")) + scale_size(range = c(0.4, 12))#change point size
```

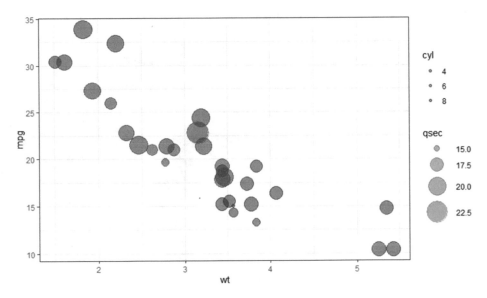

Figure 16-9. *Bubble chart mpg vs. weight for mtcars dataset*

---

[4]Yan Holtz, "Bubble Plot with Ggplot2," accessed on January 30, 2021, `www.r-graph-gallery.com/320-the-basis-of-bubble-plot.html`.

16.2.8 Scatterplot

Scatterplots show the relationship between two numerical vectors. Figure 16-10 shows a basic scatterplot.

```
unemployment = read.csv("http://datasets.flowingdata.com/unemployment-
rate-1948-2010.csv", sep = ",")
plot(1:length(unemployment$Value), unemployment$Value)
```

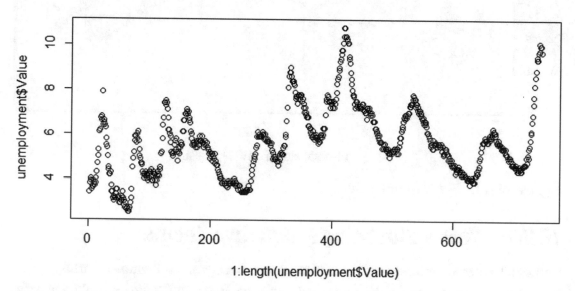

Figure 16-10. *Simple scatterplot*

16.2.9 Scatterplot with Fitted Loess Curve

Figure 16-11 applies a loess curve (local smoothing) to a scatterplot.[5]

```
scatter.smooth(x=1:length(unemployment$Value),
  y = unemployment$Value)
scatter.smooth(x= 1:length(unemployment$Value),
```

---

[5]Nathan Yau, "FlowingData," FlowingData, accessed on January 25, 2021, https://flowingdata.
 com/.

```
y = unemployment$Value, ylim=c(0,11), degree=2,
col = "#CCCCCC", span = 0.5)
```

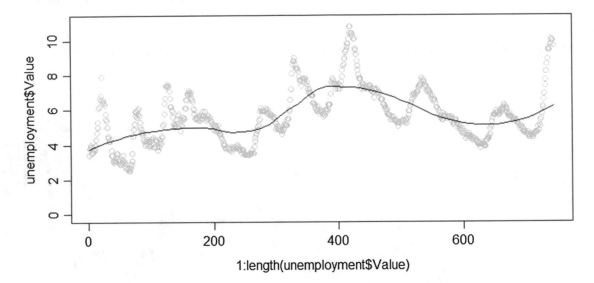

Figure 16-11. *Fitted loess curve*

16.2.10 Yearly Plot Using Alternative Theme

Ggthemr[6] has some interesting themes for ggplots. Backgrounds include colorful themes such as grape, grass, light, lilac, pale, and sea. In the following example, a simple compounded savings dataframe is created and then plotted using ggplot, as shown in Figure 16-12. Although in this case the calculations are financially exact, in many cases, you may be plotting something with some level of error or uncertainty, so I left in the automatic error bands.

```
if (!require("devtools")) install.packages("devtools")
library(devtools)
devtools::install_github('cttobin/ggthemr')
library(ggthemr)
```

---

[6]"The Ggthemr Package – Theme and Colour Your Ggplot Figures | Shane Lynn," accessed on January 26, 2021, www.shanelynn.ie/themes-and-colours-for-r-ggplots-with-ggthemr/.

```
library(lubridate)
dollars <- rep(1:1,20)   #create a vector of 20 elements
dollars[1] <- 100   # initial deposit of $100
for (i in 2:20) dollars[i] <- dollars[i -1] * 1.05
#5% annual compounded growth years 2-20
dollars <- as.data.frame(dollars)
#now add a date to x
dollars$date <- seq(as.Date("2020/1/1"),
  as.Date("2039/1/1"), "years")
names(dollars) <- c("Value", "Interest.Date")
ggthemr("earth", type="outer", layout="scientific", spacing=2)
ggplot(dollars, aes(Interest.Date, Value)) +
  geom_line() +
  geom_smooth(method='lm') +
  ggtitle("Investment value of $100 at 5% annual compounding")

## `geom_smooth()` using formula 'y ~ x'
```

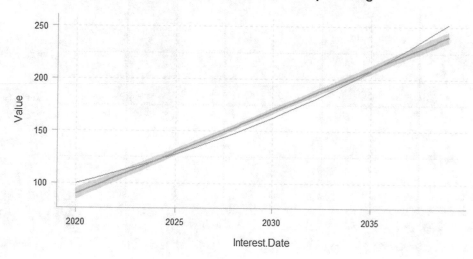

Figure 16-12. *ggplot using ggthemr theme*

16.2.11 Plot All Variables in a Dataframe Against All Other Variables

The package GGally provides a convenient, faceted series of charts showing the numerical relationships between numeric variables in a dataset. Correlation values are included. Figure 16-13 is an example, using the iris built-in dataset. Note: Mac users may need to separately insert "library(GGally)" in code before using it.

```
if (!require("GGally")) install.packages("GGally")
```

```
## Loading required package: GGally
## Registered S3 method overwritten by 'GGally':
##    method from
##    +.gg    ggplot2
```

```
ggpairs(iris)
```

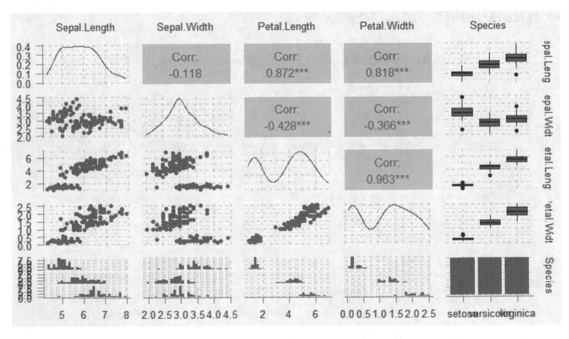

Figure 16-13. *GGally shows relationship between numeric variables*

16.2.12 Line Plot Using Two Sources of Data

Figure 16-14 is a line plot using two data sources. Both do not have to extend across the entire X axis. In the following code, note the "theme_set" function. Occasionally, the theme from one ggplot run will carry over to the next. This ensures a reset to a plain gray background.

```
n <- 20
x1 <- 1:n
y1 <- rnorm(n,0,.5)
df1 <- data.frame(x1,y1)
x2 <- (.5 * n):((1.5*n) -1)
y2 <- rnorm(n,1,.5)
df2 <- data.frame(x2,y2)
theme_set(theme_grey())
ggplot() +
    geom_line(data = df1, aes(x1,y1), color = "darkblue") +
    geom_line(data = df2, aes(x2, y2), linetype = "dashed",
    color = "red") +
    ggtitle("Two data sources plotted")
```

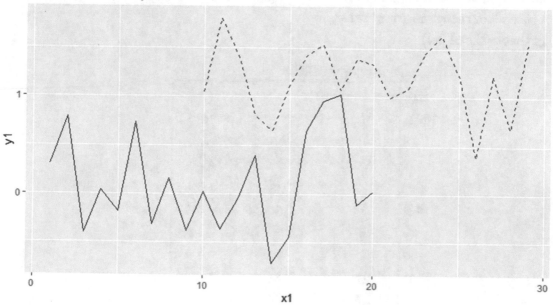

Figure 16-14. *ggplot using two data sources*

16.2.13 Correlogram

A correlogram is another way to quickly show how numeric variables in a dataset correlate with each other. Figure 16-15 is adapted from the excellent website http://r-statistics.co/Top50-Ggplot2-Visualizations-MasterList-R-Code.html.

```
library(ggplot2)
library(ggcorrplot)
```

Correlation matrix:

```
data(mtcars)
corr <- round(cor(mtcars), 1)
```

Plot:

```
ggcorrplot(corr, hc.order = TRUE,
  type = "lower",
  lab = TRUE,
  lab_size = 3,
  method="circle",
  colors = c("tomato2", "white", "springgreen3"),
  title="Correlogram of mtcars",
  ggtheme=theme_bw)
```

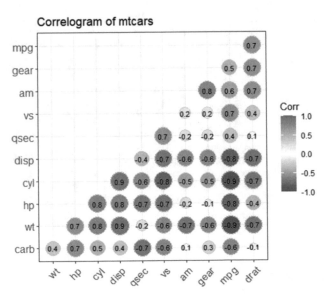

Figure 16-15. *Correlogram showing correlations between numeric variables*

16.2.14 Word Cloud

Word clouds count the frequency of each word in a group of words ("corpus") and create a colorful graphic, with the highest-frequency words having the largest graphic. The text entered should be free of quote marks and certain other confusing symbols not typically found in standard text. Figure 16-16 shows a very simple example using sample text from Wikipedia's article on football.

```
require(tm)
require(wordcloud)

## Loading required package: wordcloud
## Loading required package: RColorBrewer

wordcloud(

  "Football From Wikipedia, the free encyclopedia
Jump to navigationJump to search
This article is about the overall concept of games called football. For the
balls themselves, see Football (ball). For specific versions of the game
and other uses of the term, see Football (disambiguation).Several codes of
football. Top to bottom, left to right: association, gridiron, Australian
rules, rugby union, rugby league and Gaelic
Football is a family of team sports that involve, to varying degrees,
kicking a ball to score a goal. Unqualified, the word football normally
means the form of football that is the most popular where the word is
used. Sports commonly called football include association football (known
as soccer in some countries); gridiron football (specifically American
football or Canadian football); Australian rules football; rugby football
(either rugby union or rugby league); and Gaelic football.[1][2] These
various forms of football share to varying extent common origins and are
known as football codes.
```

There are a number of references to traditional, ancient, or prehistoric ball games played in many different parts of the world.[3][4][5] Contemporary codes of football can be traced back to the codification of these games at English public schools during the 19th century.[6][7] The expansion and cultural influence of the British Empire allowed these rules of football to spread to areas of British influence outside the directly controlled Empire.[8] By the end of the 19th century, distinct regional codes were already developing: Gaelic football, for example, deliberately incorporated the rules of local traditional football games in order to maintain their heritage.[9] In 1888, The Football League was founded in England, becoming the first of many professional football competitions. During the 20th century, several of the various kinds of football grew to become some of the most popular team sports in the world.[10]

Contents

] " ,
random.order=FALSE)

Warning in tm_map.SimpleCorpus(corpus, tm::removePunctuation):

Transformation:

```
## drops documents
## Warning in tm_map.SimpleCorpus(corpus, function(x) tm::removeWords(x,
## tm::stopwords())): transformation drops documents
```

Figure 16-16. *Word cloud using example text from Wikipedia*

16.2.15 ggplot Time Series: Airline Crash Historical Data

Figure 16-17 is a ggplot using a time series to show world airline crash fatalities since 1908.[7] Each dot represents a single crash. The data source is `https://data.world/data-society/airplane-crashes`. Data World may require registration but as of this writing does not charge a fee for airplane-crashes dataset access. The csv, as downloaded, had a character date for the year, so I converted it to an R date by using Lubridate's mdy function. I also changed the theme to the slightly clearer linedraw theme. Note in the output there were 12 rows with a missing value. That's typical of historical data of that duration.

```
setwd("h:/t")
airline.crashes <-
  read_csv("Airplane_Crashes_and_Fatalities_Since_1908.csv")
```

---

[7]"Airplane Crashes – Dataset by Data-Society," data.world, accessed on January 30, 2021, https://data.world/data-society/airplane-crashes.

```
date.and.fatalities.only <- select(airline.crashes, Date,
  Fatalities)
date.and.fatalities.only <- mutate(date.and.fatalities.only,
  crash.date = mdy(Date))

p <- ggplot(data = date.and.fatalities.only, aes(x = crash.date, y =
Fatalities)) +
  geom_point() +
  labs(x = "Date",
    y = "Airline Fatalities",
    title = "Flying is safer than driving but you could still win the
devil's lottery",
      subtitle = "source = https://data.world/data-
    society/airplane-crashes")

p + theme_linedraw()
```

Warning: Removed 12 rows containing missing values (geom_point).

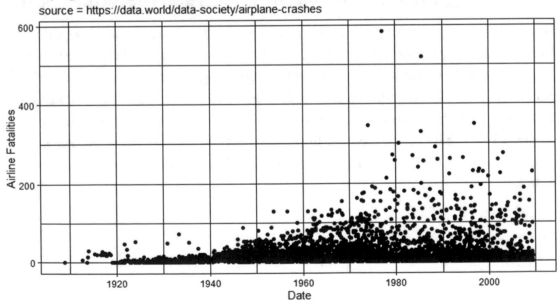

Figure 16-17. Fatalities per airline crash, since 1908

16.2.16 Textplot

In 2020, a new package, textplot, was added to CRAN. It provides handy functionality for plotting text values, showing correlations and frequencies:

- Text cooccurrences

- Text clusters (in casu biterm clusters)

- Dependency parsing results

- Text correlations and text frequencies

Figure 16-18 shows the output of one of textplot's bar graphs[8]:

```
if (!require("textplot")) install.packages("textplot")
if (!require("udpipe")) install.packages("udpipe")
data(brussels_listings, package = 'udpipe')
x <- table(brussels_listings$neighbourhood)
x <- sort(x)
textplot_bar(x,
panel = "Locations", col.panel = "darkgrey", xlab = "Listings",
cextext = 0.75, addpct = TRUE, cexpct = 0.5)
```

---

[8]https://cran.r-project.org/web/packages/textplot/textplot.pdf. Accessed on January 25, 2021.

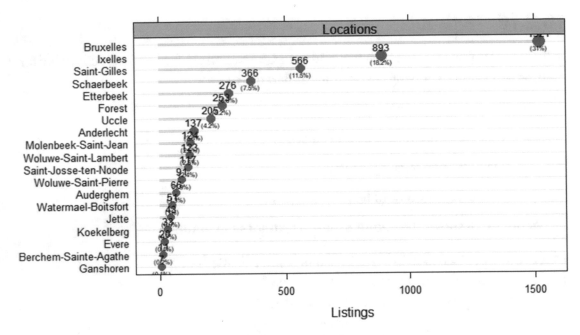

Figure 16-18. *Textplot of listings by location in Brussels*

16.2.17 Dot Plot

Dot plot has been around for years. It is a favorite of magazines and easy to code. Figure 16-19 shows example output using the mpg built-in dataset.

```
library(tidyverse)
theme_set(theme_classic())
ggplot(mpg, aes(x=manufacturer, y=cty)) + geom_point(col="tomato2", size=3) +
  geom_segment(aes(x=manufacturer, xend=manufacturer,
  y=min(cty), yend=max(cty)), linetype = "dashed", size=0.1) +
  labs(title="Dot Plot",
  subtitle="Manufacturer(Make) Vs Mileage in the City", caption="source:
  mpg") + coord_flip()
```

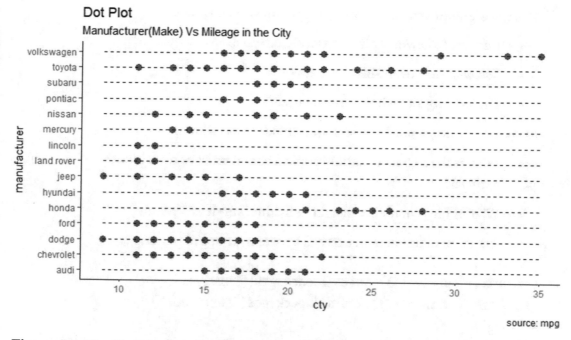

Figure 16-19. *Simple dot plot showing city mileage vs. car manufacturer*

16.2.18 Survival Analysis

In medical research, a standard question is to ask how long a patient will survive after a given treatment/nontreatment for a disease. Historically, survival analysis has been of keen interest to insurance companies, as their actuaries refine rates based on historical data. R has some powerful tools to address these questions. Over time, survival analysis has come to include more than medical/insurance topics. A business with rapid customer turnover will want to know how much to invest in acquiring a new customer (advertising costs, etc.) based on the estimated "lifetime" of that customer—until the customer drops one company and signs up with another. Figure 16-20 shows a typical graphic from the survival package.

The website sdhta.com lists example uses for survival analysis:[9]

- *Cancer studies* for patient survival time analyses

- *Sociology* for "event-history analysis"

- *Engineering* for "failure-time analysis"

In cancer studies, typical research questions are as follows:

- What is the impact of certain clinical characteristics on patient's survival?

- What is the probability that an individual survives 3 years?

- Are there differences in survival between groups of patients?

```
if (!require("survival")) install.packages("survival")
if (!require("survminer")) install.packages("survminer")
data("lung")
head(lung)
```

```
##    inst time status age sex ph.ecog ph.karno pat.karno meal.cal wt.loss
## 1     3  306      2  74   1       1       90       100     1175      NA
## 2     3  455      2  68   1       0       90        90     1225      15
## 3     3 1010      1  56   1       0       90        90       NA      15
## 4     5  210      2  57   1       1       90        60     1150      11
## 5     1  883      2  60   1       0      100        90       NA       0
## 6    12 1022      1  74   1       1       50        80      513       0
```

Fit a model (kaplan-Meier survival estimate):

```
fit <- survfit(Surv(time, status) ~ sex, data = lung)
ggsurvplot(fit,
           pval = TRUE, conf.int = TRUE,
           risk.table = TRUE, # Add risk table
           risk.table.col = "strata", # Change risk table color by groups
           linetype = "strata", # Change line type by groups
```

---

[9]"Survival Analysis Basics – Easy Guides – Wiki – STHDA," accessed on January 26, 2021, www.sthda.com/english/wiki/survival-analysis-basics.

```
surv.median.line = "hv", # Specify median survival
ggtheme = theme_bw(), # Change ggplot2 theme
palette = c("#E7B800", "#2E9FDF"))
```

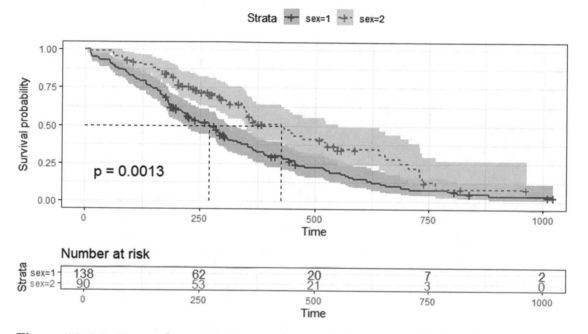

Figure 16-20. *Example survival curve (www.sthda.com/english/wiki/ survival-analysis-basics)*

## CHAPTER 17

Simple Prediction Methods

Predictive modeling has gotten sophisticated over the years. It is a major discipline of data science and includes some of the most sophisticated mathematics and statistical engineering found anywhere. That being said, some of the tools for prediction are straightforward. We can start with the package prophet, graciously given to the world by Facebook. If you compare prophet to some of the older time series prediction methods, you will be impressed with its simplicity. It has all the trending, seasonality, and other mathematical patterns of older systems but does not force you to "get involved." You can just enter a dataframe with one column of dates and another column of some numeric value (birds per square mile, prisoners in Alabama, rain forest size, etc.) and then predict future values. The most time-consuming part is getting the dates and column headers in the format prophet requires. After prophet, I'll present an older method, Holt-Winters (time series) and multivariate regression, where one or more variables (*predictors*) are used to estimate some variable of interest, termed the *response*.

17.1 Prophet: Time Series Modeling

Prophet does a remarkable job of hiding complexity so that all you have to do is get two columns in a dataframe: a date and a numeric variable of interest (what you are trying to predict). Unfortunately, there is no compiled version of prophet on Mac.

```
library(prophet)
library(lubridate)
```

W. Yarberry, *CRAN Recipes*, https://doi.org/10.1007/978-1-4842-6876-6_17

Start by creating an artificial time series dataset, starting with the first day in 2001:

```
start1 <- mdy("01/01/2001")
end1 <- mdy("01/01/2003")
the.days <- seq(start1, end1,"days") #daily
```

Use a handy built-in dataset, co2 (that's c-alpha o-2), which relates to co2 uptake in plants:

```
co2.levels <- as.numeric(co2)
max.row <- length(co2.levels)
the.days <- the.days[1:max.row]
df <- data.frame(the.days = the.days, co2.levels = co2.levels)
names(df) <- c("ds", "y")
```

The dataframe column names of ds and y are required. It will not work with any other names. The above dataframe construction works better than cbind in this case. cbind can change the format on dates. Sometimes R just does not do what you would expect.

```
tail(df)
```

```
##                  ds      y
## 463 2002-04-08 364.52
## 464 2002-04-09 362.57
## 465 2002-04-10 360.24
## 466 2002-04-11 360.83
## 467 2002-04-12 362.49
## 468 2002-04-13 364.34
```

```
m <- prophet(df)
```

Predict the value of co2 for 50 periods (days in this case) into the future:

```
future <- make_future_dataframe(m, periods = 50)
tail(future)
```

```
##                  ds
## 513 2002-05-28
## 514 2002-05-29
## 515 2002-05-30
```

```
## 516 2002-05-31
## 517 2002-06-01
## 518 2002-06-02

forecast <- predict(m, future)
tail(forecast[c('ds', 'yhat', 'yhat_lower', 'yhat_upper')])

##              ds      yhat yhat_lower yhat_upper
## 513 2002-05-28 369.3050    366.4695   371.8820
## 514 2002-05-29 369.4339    366.8909   372.0976
## 515 2002-05-30 369.5444    366.9370   372.1789
## 516 2002-05-31 369.6570    367.0379   372.3523
## 517 2002-06-01 369.8023    367.1528   372.4809
## 518 2002-06-02 369.8654    366.9304   372.5212

plot(m, forecast)
```

Figure 17-1 shows a prediction graph from the prophet system.

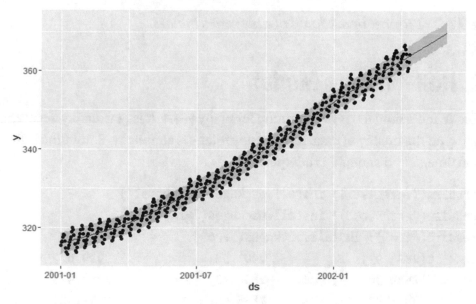

Figure 17-1. *Prophet graph predicting 50 days into the future*

Figure 17-2 breaks down a time series into factors.

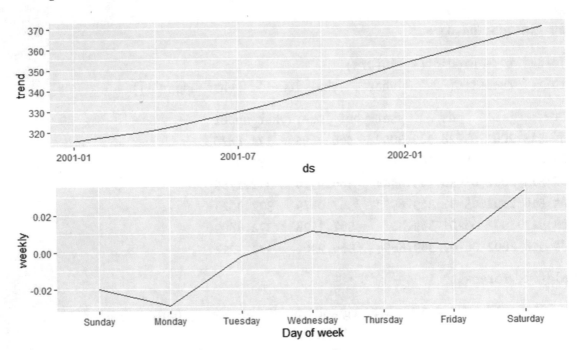

Figure 17-2. *Prophet breakdown of time series factors*

17.2 Holt-Winters Model

The Holt-Winters model has been around for many years. It is reasonably accurate, depending on the quality of data. Its great advantage is simplicity. Sometimes, given available time, "good enough" is adequate.

```
if (!require("forecast")) install.packages("forecast")
if (!require("graphics")) install.packages("graphics")
if (!require("zoo")) install.packages("zoo")
values1 <- c(2821.93,        2901.52,        2929.67,
             2839.96,        2896.72,        2919.37,
             2818.37,        2888.60,        2915.56,
             2840.69,        2878.05,        2905.97,
             2850.13,        2871.68,        2914.00,
             2857.05,        2877.13,        2913.98,
             2862.96,        2887.89,        2924.59,
```

2861.82,	2888.92,	2923.43,
2856.98,	2904.18,	2925.51,
2874.69,	2904.98,	2901.61,
2896.74,	2888.80,	2885.57,
2897.52,	2904.31,	2884.43,
2914.04,	2907.95,	2880.34,
2901.13,	2930.75,	2785.68,
2728.37,	2755.88,	2711.74,
2767.13,	2740.69,	2999.91,
2750.79,	2656.10,	3013.77,
2809.92,	2705.57,	3014.30,
2809.21,	2658.69,	3004.04,
2768.78,	2641.25,	2984.42,
2767.78,	2682.63,	2995.11)

Create a time series of monthly data, starting December 2009:

```
values.as.time.series <- ts(values1, start = c(2009,12),
  frequency = 12)

model.hw = HoltWinters(values.as.time.series)

p = predict(model.hw, 15) #look at 15 periods into the future
p

##              Jan      Feb      Mar      Apr      May      Jun
Jul      Aug
## 2015                            2991.145 2984.838 2976.383 2996.194 2991.383
                                   2983.245
## 2016 3026.521 3027.701 3023.736 3017.429 3008.974
##           Sep      Oct      Nov      Dec
## 2015 2997.872 3018.700 3019.718 3037.716
## 2016

m = merge(a = as.zoo(values.as.time.series),
   b = as.zoo(p))
m = as.ts(m)
ts.plot(m, ylab = "Values")
```

Figure 17-3 uses an older algorithm, Holt-Winters, to predict future values.

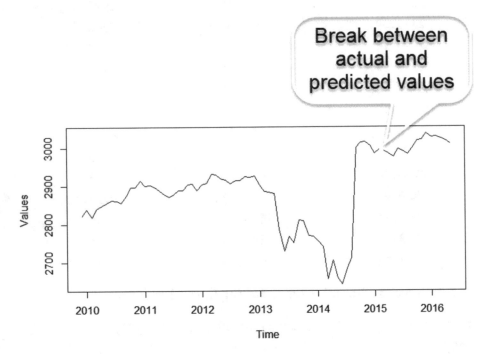

Figure 17-3. *Holt-Winters time series prediction algorithm*

17.3 Multivariate Regression

In multivariate regression, one or more predictors are read by a model which then estimates a response (the variable of interest). Selecting predictors is an art and a science. If you pick too many, then the algorithm will "overfit" to specific historical data and will not generalize to the future. If you pick too few, the predictive power will suffer. The following example is a bare-bones predictive model of miles per gallon based on various engine parameters. The "fit" model is created using the predictors of am, vs, cyl, disp, hp, wt, and drat. The model shown here could be improved by further analysis.

```
data(mtcars)

fit <- lm(mpg ~ am + vs + cyl + disp + hp + wt + drat, data = mtcars)
car.test <- mtcars[c(1,5,15,21),]  #test with four records
my.prediction <- fit %>% predict(car.test)
car.test[,1]
```

```
## [1] 21.0 18.7 10.4 21.5
```

```
my.prediction
```

```
##          Mazda RX4   Hornet Sportabout  Cadillac Fleetwood
           Toyota Corona
##          23.02141            17.44326            12.04897
           23.99681
```

The model is reasonably close. For example, it predicts 23 miles per gallon vs. the actual mpg of 21 for the Mazda RX4.

## CHAPTER 18

Smorgasbord of Simple Statistical Tests

The following are a sampling of one-liner tests. They provide a wealth of information on any numeric series/column with little code required. If you are doing upfront data exploration, put a chunk of these at the beginning of your program.

18.1 Basic Numeric Vector Tests

For simplicity, let's use the built-in dataset "Animals." Using brain size as a numeric variable, the following code provides plenty of descriptive information. Figure 18-1, a histogram of brain weights, shows that most are less than 1000 ccs.

```
library(datasets)
library(Hmisc)
library(MASS)
data("Animals")
A <- Animals
median(A$brain) # "in the middle"

## [1] 137

mean(A$brain)   #average

## [1] 574.5214

range(A$brain)   #min and max

## [1]    0.4 5712.0
```

```
mode(A$brain)
```

```
## [1] "numeric"
```

```
sd(A$brain) #standard deviation
```

```
## [1] 1334.929
```

```
plot(A$brain)
```

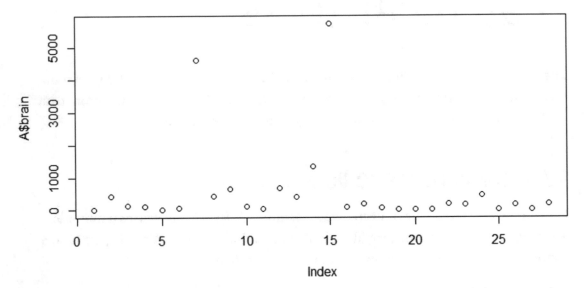

Figure 18-1. *Plot shows just a few brain weight outliers*

Is this a normally distributed population? No. Figure 18-2 shows a decidedly non-straight line, so the underlying population is not normally distributed.

```
qqnorm(A$brain)
```

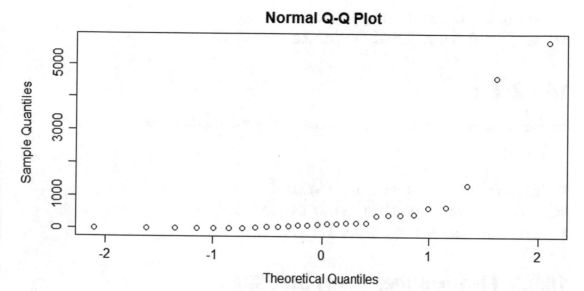

Figure 18-2. *Normal Q-Q plot, indicating a non-normal population in this case*

18.1.1 Coefficient of Variation

I've always thought the coefficient of variation should get more respect. It is independent of dimension (dimensionless) so it permits comparison of disparate populations. For example, assume you are awarding bonuses to factory managers. One manager manages hammer production and the other sewing pin production. You want both products to be as uniform as possible. You certainly cannot use standard deviation of weight as a way to compare performance; otherwise, the hammer manager would always lose relative to the pin manager. The fair way to compare each manager is by coefficient of variation.

```
sd(A$brain)/median(A$brain)      #coefficient of variation
```

```
## [1] 9.744009
summary(A)
```

```
##         body              brain
##  Min.   :    0.02   Min.   :   0.40
##  1st Qu.:    3.10   1st Qu.:  22.23
##  Median :   53.83   Median : 137.00
##  Mean   : 4278.44   Mean   : 574.52
```

```
##   3rd Qu.:   479.00    3rd Qu.:  420.00
##   Max.     :87000.00   Max.     :5712.00
```

18.1.2 Str

Str is another convenience summary function. Think of it as "structure."

```
Str(A)
```

```
## 'data.frame':     28 obs. of  2 variables:
##  $ body : num   1.35 465 36.33 27.66 1.04 ...
##  $ brain: num   8.1 423 119.5 115 5.5 ...
```

18.1.3 Five-Number Tukey Statistics

The five-number Tukey statistic is always handy. It includes the minimum, lower hinge, median, upper hinge, and maximum.

```
fivenum(A$brain)
```

```
## [1]    0.40    18.85   137.00   421.00 5712.00
```

```
describe(A$brain)
```

```
## A$brain
##          n   missing distinct     Info     Mean      Gmd      .05      .10
##         28         0       27        1    574.5    928.6    1.315    2.670
##        .25       .50      .75      .90      .95
##     22.225   137.000  420.000  872.000 3453.950
##
## lowest :    0.4     1.0     1.9     3.0     5.5, highest:  655.0
##    680.0 1320.0 4603.0 5712.0
##
## Value         0    10    30    50    60    70   120   150   160
##    180    410
## Frequency     4     3     1     1     1     1     3     1     1
##     3     1
```

```
## Proportion 0.143 0.107 0.036
0.036 0.036 0.036 0.107 0.036 0.036 0.107 0.036
##
## Value          420    440    660
680   1320   4600   5710
## Frequency        2      1      1
1      1      1      1
## Proportion 0.071 0.036 0.036
0.036 0.036 0.036 0.036
```

Key statistics for fivenum:

- n, nmiss, unique, mean, 5,10,25,50,75,90,95th percentiles.

- 5 lowest and 5 highest scores.

- At a brain weight of 872, animal is in the 90th percentile.

18.2 Attractive Tabular Output

18.2.1 sjPlot

The sjPlot package provides an attractive tabular wrapper, as shown in Table 18-1. One of R's strengths is the availability of a package/function for nearly all common presentation formats.

```
tab_df(mpg[1:10,]) #uses built-in mpg dataframe
```

Table 18-1. *Tabular presentation in sjPlot package*

manufacturer	model	displ	year	cyl	trans	drv	cty	hwy	fl	class
audi	a4	1.80	1999	4	auto(l5)	f	18	29	p	compact
audi	a4	1.80	1999	4	manual(m5)	f	21	29	p	compact
audi	a4	2.00	2008	4	manual(m6)	f	20	31	p	compact
audi	a4	2.00	2008	4	auto(av)	f	21	30	p	compact
audi	a4	2.80	1999	6	auto(l5)	f	16	26	p	compact
audi	a4	2.80	1999	6	manual(m5)	f	18	26	p	compact
audi	a4	3.10	2008	6	auto(av)	f	18	27	p	compact
audi	a4 quattro	1.80	1999	4	manual(m5)	4	18	26	p	compact
audi	a4 quattro	1.80	1999	4	auto(l5)	4	16	25	p	compact
audi	a4 quattro	2.00	2008	4	manual(m6)	4	20	28	p	compact

18.2.2 Formattable

Table 18-2 shows another cosmetic option, using the formattable package.

```
if (!require("formattable")) install.packages("formattable")
```

```
## Loading required package: formattable
##
## Attaching package: 'formattable'
## The following object is masked from 'package:MASS':
##
##      area
```

```
library(formattable)
formattable (mpg[1:10,])
```

Table 18-2. *Partial copy of the mpg dataset with enhanced appearance by* *formattable*

manufacturer	model	displ	year	cyl	trans	drv	cty	hwy	fl	class
audi	a4	1.8	1999	4	auto(l5)	f	18	29	p	compact
audi	a4	1.8	1999	4	manual(m5)	f	21	29	p	compact
audi	a4	2.0	2008	4	manual(m6)	f	20	31	p	compact
audi	a4	2.0	2008	4	auto(av)	f	21	30	p	compact
audi	a4	2.8	1999	6	auto(l5)	f	16	26	p	compact
audi	a4	2.8	1999	6	manual(m5)	f	18	26	p	compact
audi	a4	3.1	2008	6	auto(av)	f	18	27	p	compact
audi	a4 quattro	1.8	1999	4	manual(m5)	4	18	26	p	compact
audi	a4 quattro	1.8	1999	4	auto(l5)	4	16	25	p	compact –
audi	a4 quattro	2.0	2008	4	manual(m6)	4	20	28	p	compact

18.3 Distributions

18.3.1 Normal

Create five normally distributed random numbers as a numeric vector:

```
x <- rnorm(n=5)
x  # five normally distributed random numbers (numeric, vector)

## [1]   0.234   0.208   1.415  -1.218  -1.270
```

Twenty normal random numbers with mean and standard deviation specified:

```
x <- rnorm(n=20, mean=1000, sd=100)
x

## [1]   960 1056 1007 1092  958  961  797 1083 1030 1084  910 1143 1042
1012 1074
## [16]  906  992  860 1073 1021
```

What value is required to be in the 95th percentile, given a specified mean and standard deviation?

```
x <- qnorm(0.95, mean=100, sd=15)
x
```

```
## [1] 125
```

What is the minimum IQ to get into Mensa (top 2%)?

```
x <- qnorm(0.98, mean = 100, sd = 15)
x
```

```
## [1] 131
```

Given a mean of 17.47 grams and standard deviation of 28.1 grams, what is the probability that a sample widget will weigh more than 19 grams?

```
x <- 1 - pnorm(19, mean=17.46, sd=18.1)
x
```

```
## [1] 0.466
```

18.3.2 Binomial

What is the probability of six successes out of ten, assuming a 30% success rate?

```
x <- dbinom(x = 6, size = 10, prob = .3)
x
```

```
## [1] 0.0368
```

18.3.3 Poisson

Example use for the Poisson distribution functions: If there are 12 cars crossing a bridge per minute on average, find the probability of having 17 or more cars crossing the bridge in a particular minute.

The probability of having 16 or *fewer* cars crossing the bridge in a particular minute is given by the function ppois.

```
ppois(16, lambda=12)    # lower tail
```

```
## [1] 0.899
```

The probability of having 17 or *more* cars crossing the bridge in a minute is in the upper tail of the probability density function.

```
ppois(16, lambda=12, lower=FALSE)    # upper tail
```

```
## [1] 0.101
```

18.4 Quick Data Exploration
18.4.1 Convenience Summaries
18.4.11 dfSummary

As an FYI, dfSummary is probably more trouble than it is worth on Mac; for example, you must install XQuartz for it to work. Probably best to use other summary functions for Mac.

```
library(summarytools)
dfSummary(airquality)
```

```
## Data Frame Summary
## airquality
## Dimensions: 153 x 6
## Duplicates: 0
##
## ------------------------------------------------------------------
----------------## No   Variable     Stats / Values                Freqs (% of
Valid)    Graph                       Valid     Missing
## ----  ----------- ------------------------ --------------------- -------
---------------- ---------- ---------
## 1     Ozone        Mean (sd) : 42.1 (33)       67 distinct values    : .
116           37
```

```
##        [integer]   min < med < max:                          : :
(75.8%)    (24.2%)
##                    1 < 31.5 < 168                            : :
##                    IQR (CV) : 45.2 (0.8)                     : : : :
##                                                              : : : :
: . .
##
## 2    Solar.R    Mean (sd) : 186 (90.1)    117 distinct values
:             146         7
##        [integer]   min < med < max:
: :              (95.4%)    (4.6%)
##                    7 < 205 < 334                             . . . .
: :
##                    IQR (CV) : 143 (0.5)                      : : : :
: :
##                                                              : : : :
: : :
##
## 3    Wind       Mean (sd) : 10 (3.5)      31 distinct values        :
153        0
##        [numeric]   min < med < max:                                 :
:               (100.0%)   (0.0%)
##                    1.7 < 9.7 < 20.7                          . :
: : :
##                    IQR (CV) : 4.1 (0.4)                      : :
: : :
##                                                              . : : :
: : : :   .
##
## 4    Temp       Mean (sd) : 77.9 (9.5)    40 distinct
values           : :              153        0
##        [integer]   min < med < max:
: :              (100.0%)   (0.0%)
##                    56 < 79 < 97                                     :
: : :
```

```
##                      IQR (CV) : 13 (0.1)                           . : :
: : : :
##                                                                  : : : :
: : : : .
##
## 5      Month       Mean (sd) : 7 (1.4)        5 : 31 (20.3%)        IIII
153           0
##         [integer]   min < med < max:          6 : 30 (19.6%)        III
(100.0%)    (0.0%)
##                      5 < 7 < 9                 7 : 31 (20.3%)        IIII
##                      IQR (CV) : 2 (0.2)        8 : 31 (20.3%)        IIII
##                                                9 : 30 (19.6%)        III
##
## 6      Day         Mean (sd) : 15.8 (8.9)    31 distinct values     : : : :
: :               153          0
##         [integer]   min < med < max:                                : : : :
: :               (100.0%)    (0.0%)
##                      1 < 16 < 31                                     : : : :
: :
##                      IQR (CV) : 15 (0.6)                             : : : :
: :
##                                                                     : : : :
: : .
## -----------------------------------------------------------------------------
```

Although dfSummary provides a wealth of detail, the output is cosmetically awkward. I suppose if you work with it long enough, you get used to it. It reminds me of an individual working as a programmer in the administrative group of my college. He created reports with no headings. I asked him how he would feel if he got something in the mail with nothing but numbers on a page. He said "no problem, they'll get used to it."

18.4.12 Skimr

Skimr is another cosmetically bare bones but information-rich summary function:

```
library(skimr)
skim(airquality)
```

Data summary
Name airquality
Number of rows 153
Number of columns 6

Column type frequency:
numeric 6

Group variables None
Variable type: numeric

skim_variable	n_missing	complete_rate	mean	sd	p0	p25	p50	p75	p100	hist
Ozone	37	0.76	42.13	32.99	1.0	18.0	31.5	63.2	168.0	▂▁__
Solar.R	7	0.95	185.93	90.06	7.0	115.8	205.0	258.8	334.0	▂▃█
Wind	0	1.00	9.96	3.52	1.7	7.4	9.7	11.5	20.7	▂█
Temp	0	1.00	77.88	9.47	56.0	72.0	79.0	85.0	97.0	▂█▃
Month	0	1.00	6.99	1.42	5.0	6.0	7.0	8.0	9.0	██
Day	0	1.00	15.80	8.86	1.0	8.0	16.0	23.0	31.0	██

```
# Data summary
# Name
# Number of rows
# Number of columns
# _____
```

```
# Column type frequency:
# numeric
# _____
# Group variables
# Variable type: numeric
# skim_
# variable
# Ozone
# Solar.R
# Wind
# Temp
# Month
# Day
```

18.4.13 Hmisc

In the following code, I use the syntax "package.name::function." The function "describe" is used by other packages, so using an explicit statement of the package from which the function is obtained eliminates ambiguity.

```
Hmisc::describe(airquality)
```

```
## airquality
##
## 6  Variables      153  Observations
## --------------------------------------------------------------------
## Ozone
##         n  missing distinct    Info    Mean     Gmd     .05     .10
##       116       37       67   0.999   42.13   35.28    7.75   11.00
##       .25      .50      .75     .90     .95
##     18.00    31.50    63.25   87.00  108.50
##
## lowest :   1    4    6    7    8, highest: 115 118 122 135 168
## --------------------------------------------------------------------
## Solar.R
##         n  missing distinct    Info    Mean     Gmd     .05     .10
##       146        7      117       1   185.9   102.7   24.25   47.50
```

```
##      .25       .50       .75       .90       .95
##   115.75    205.00    258.75    288.50    311.50
##
## lowest :   7   8   13   14   19, highest: 320 322 323 332 334
## ----------------------------------------------------------------------
## Wind
##          n  missing distinct      Info      Mean       Gmd       .05       .10
##        153        0       31     0.997     9.958     3.964      4.60      5.82
##      .25       .50       .75       .90       .95
##     7.40      9.70     11.50     14.90     15.50
##
## lowest :   1.7   2.3   2.8   3.4   4.0, highest: 16.1 16.6 18.4 20.1 20.7
## ----------------------------------------------------------------------
## Temp
##          n  missing distinct      Info      Mean       Gmd       .05       .10
##        153        0       40     0.999     77.88     10.74      60.2      64.2
##      .25       .50       .75       .90       .95
##     72.0      79.0      85.0      90.0      92.0
##
## lowest : 56 57 58 59 61, highest: 92 93 94 96 97
## ----------------------------------------------------------------------
## Month
##          n  missing distinct      Info      Mean       Gmd
##        153        0        5      0.96     6.993     1.608
##
## lowest : 5 6 7 8 9, highest: 5 6 7 8 9
##
## Value          5      6      7      8      9
## Frequency     31     30     31     31     30
## Proportion 0.203  0.196  0.203  0.203  0.196
## ----------------------------------------------------------------------
## Day
##          n  missing distinct      Info      Mean       Gmd       .05       .10
##        153        0       31     0.999      15.8     10.26       2.0       4.0
##      .25       .50       .75       .90       .95
```

```
##      8.0      16.0      23.0      28.0      29.4
##
## lowest :  1  2  3  4  5, highest: 27 28 29 30 31
## -------------------------------------------------------------------
```

18.4.14 Summary

```
summary(airquality)
```

```
##      Ozone            Solar.R          Wind            Temp            Month
## Min.   :  1.0    Min.   :  7     Min.   : 1.70    Min.   :56.0    Min.   :5.00
## 1st Qu.: 18.0    1st Qu.:116     1st Qu.: 7.40    1st Qu.:72.0    1st Qu.:6.00
## Median : 31.5    Median :205     Median : 9.70    Median :79.0    Median :7.00
## Mean   : 42.1    Mean   :186     Mean   : 9.96    Mean   :77.9    Mean   :6.99
## 3rd Qu.: 63.2    3rd Qu.:259     3rd Qu.:11.50    3rd Qu.:85.0    3rd Qu.:8.00
## Max.   :168.0    Max.   :334     Max.   :20.70    Max.   :97.0    Max.   :9.00
## NA's   :37       NA's   :7
##      Day
## Min.   : 1.0
## 1st Qu.: 8.0
## Median :16.0
## Mean   :15.8
## 3rd Qu.:23.0
## Max.   :31.0
##
```

18.5 Numeric Vectors

The base R table command shows a count of each value in a vector.

```
x <- mtcars$mpg
table(x)
```

```
## x
## 10.4 13.3 14.3 14.7   15 15.2 15.5 15.8 16.4 17.3 17.8 18.1 18.7 19.2
## 19.7   21
```

```
##     2    1    1    1    1    2    1    1    1    1    1    1    1
2    1    2
## 21.4 21.5 22.8 24.4   26 27.3 30.4 32.4 33.9
##     2    1    2    1    1    1    2    1    1
```

To show the number of duplicates in a vector, subtract the number of unique values from the total count of values.

```
length(x) - length(unique(x))
```

```
## [1] 7
```

18.6 Heatmap of Correlations

Heatmaps show correlations between numeric variables, with darker colors indicating stronger relationships. Figure 18-3 is a heatmap of the built-in dataset airquality. The reshape2 package is required for this visual.

```
library(reshape2)
```

```
##
## Attaching package: 'reshape2'
## The following object is masked from 'package:tidyr':
##
##       smiths
```

```
data <- airquality[,1:4]
qplot(x=Var1, y=Var2, data=melt(cor(data, use="p")), fill=value,
geom="tile",
      main="correlations") +  scale_fill_gradient2(limits=c(-1, 1))
```

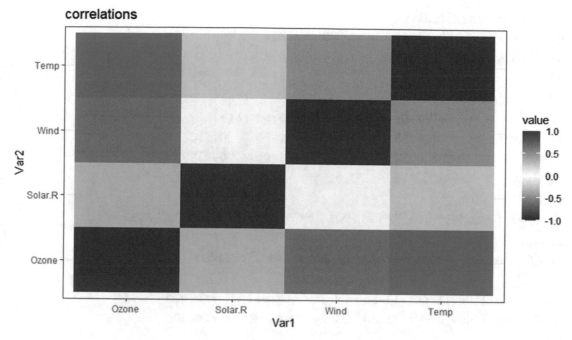

Figure 18-3. *Heatmap of correlations in airquality dataframe*

18.7 Easy Models

18.7.1 Generalized Linear Model

Using mtcars as data input, this model predicts the probability that an engine will be V shaped, given the weight and displacement specified. The function is glm, generalized linear model.

```
my.model <- glm(formula= vs ~ wt + disp, data=mtcars,
  family=binomial)
summary(my.model)

##
## Call:
## glm(formula = vs ~ wt + disp, family = binomial, data = mtcars)
##
```

```
## Deviance Residuals:
##    Min     1Q  Median     3Q     Max
## -1.675  -0.284  -0.084   0.573   2.082
##
## Coefficients:
##               Estimate Std. Error z value Pr(>|z|)
## (Intercept)    1.6086     2.4390    0.66   0.510
## wt             1.6264     1.4907    1.09   0.275
## disp          -0.0344     0.0154   -2.24   0.025 *
## ---
## Signif. codes:  0 '***' 0.001 '**' 0.01 '*' 0.05 '.' 0.1 ' ' 1
##
## (Dispersion parameter for binomial family taken to be 1)
##
##      Null deviance: 43.86  on 31  degrees of freedom
## Residual deviance: 21.40  on 29  degrees of freedom
## AIC: 27.4
##
## Number of Fisher Scoring iterations: 6
```

In this example, a weight of 2.1 and displacement of 180 is associated with a 23.6% probability that the engine is V shaped.

```
newdata = data.frame(wt = 2.1, disp = 180)
predict(my.model, newdata, type="response")
```

```
##      1
## 0.236
```

18.7.2 Prediction Using Simple Linear Regression

With the growth of machine learning and other sophisticated prediction algorithms, linear regression is looking more like yesterday's tool than anything else. Nonetheless, it can still be a useful predictor in some cases. Plus, it is easy and fast. This model uses the cars built-in dataset and predicts distance traveled as a function of speed, for various speeds. For example, if car's speed is 10, then I'll travel 21.7 feet.

```
Input_variable_speed <- data.frame(speed =
     c(10,12,15,18,10,14,20,25,14,12))
linear_model = lm(dist~speed, data = cars)
```

```
my_predictions <- predict(linear_model,
    newdata = Input_variable_speed)
my_predictions
```

```
##        1        2        3        4        5        6        7        8
## 21.74499 29.60981 41.40704 53.20426 21.74499 37.47463 61.06908 80.73112
##        9       10
## 37.47463 29.60981
```

18.7.3 Text Mining/Analytics

Text mining applications are growing rapidly, fueled by the exobytes of words generated in business, scientific research, and of course social media. Wikipedia notes the following[1]:

> *Text mining, also referred to as text data mining, similar to text analytics, is the process of deriving high-quality information from text. It involves "the discovery by computer of new, previously unknown information, by automatically extracting information from different written resources."[1] Written resources may include websites, books, emails, reviews, and articles. High-quality information is typically obtained by devising patterns and trends by means such as statistical pattern learning. According to Hotho et al. (2005) we can differ three different perspectives of text mining: information extraction, data mining, and a KDD (Knowledge Discovery in Databases) process.[2] Text mining usually involves the process of structuring the input text (usually parsing, along with the addition of some derived linguistic features and the removal of others, and subsequent insertion into a database), deriving patterns within the structured data, and finally evaluation and interpretation of the output. "High quality" in text mining usually refers to some combination of relevance, novelty, and interest. Typical text mining tasks include text categorization, text clustering, concept/entity extraction, production of granular taxonomies, sentiment analysis, document summarization, and entity relation modeling (i.e., learning relations between named entities).*

---

[1]"Text Mining," in Wikipedia, January 15, 2021, https://en.wikipedia.org/w/index.php?title=Text_mining&oldid=1000584645.

> *Text analysis involves information retrieval, lexical analysis to study word frequency distributions, pattern recognition, tagging/annotation, information extraction, data mining techniques including link and association analysis, visualization, and predictive analytics. The overarching goal is, essentially, to turn text into data for analysis, via application of natural language processing (NLP), different types of algorithms and analytical methods.*

This example uses the tm package to remove punctuation symbols and numbers from a string.

```
require(tm)
x <- c("Please....don't throw me in #1 briar #### $$$$  **** patch")

## Loading required package: tm
## Loading required package: NLP
##
## Attaching package: 'NLP'
## The following object is masked from 'package:ggplot2':
##
##       annotate

y <- removePunctuation(x)
y

## [1] "Pleasedont throw me in 1 briar     patch"

x <- c("this has both numbers like 3,5 and letters and punctuation.....")
z <- removeNumbers(x)
z

## [1] "this has both numbers like , and letters and punctuation....."
```

18.8 Sampling

18.8.1 Sampling with Replacement

If you put black and white marbles in a bag and pull one out with your eyes closed, that's sampling. If you put the marble back in the bag and pull one out again, that's sampling with replacement. R's replace option provides the functionality to sample either way.

```
total.quantity.of.samples <- 200
number.to.be.drawn <- 50
x <- sample(total.quantity.of.samples, number.to.be.drawn, replace = TRUE)
x
```

```
##  [1] 185 137 199 160 137  38  19 169  88 197  93 189  41 103  64 168
156  34 167
## [20]  76 157  18   8  44  82 197 179 161  98 165 164 170  97  25 170  87
151  46
## [39] 159 184  75   5  25 100  97 196  27  62 108 152
```

18.8.2 Split Sample (Control vs. Treatment)

A typical use for a split sample is to create a treatment and control population:

```
my.sample <- 20
y <- split(sample(my.sample), rep(c("control","treatment"),10))
y
```

```
## $control
##  [1]   3 18 14   2 13 19 10 16 17   4
##
## $treatment
##  [1] 11 12 15   6  8  5  9  7  1 20
It is often helpful to convert the list y into a dataframe:
```

```
z <- as.data.frame(y)
head(z,5)
```

```
##   control treatment
## 1      11        12
```

## 2	10	22
## 3	28	16
## 4	29	19
## 5	2	7

18.9 Financial Functions

Use Yahoo to provide stock price history and month-to-month changes for a specific stock. Figures 18-4 and 18-5 show plots of the results using the financial workhorse package quantmod.

```
require(quantmod)
price = getSymbols('T', src = 'yahoo',
    from = '2002-01-01', auto.assign = F)
getOption("getSymbols.env")
price = Cl(to.monthly(price, indexAt='endof'))
plot(price, main = "AT&T EOM Stock Price")
```

Figure 18-4. *Stock price history generated by quantmod*

```
plot(diff(price), main = "Month to Month change:AT&T")
```

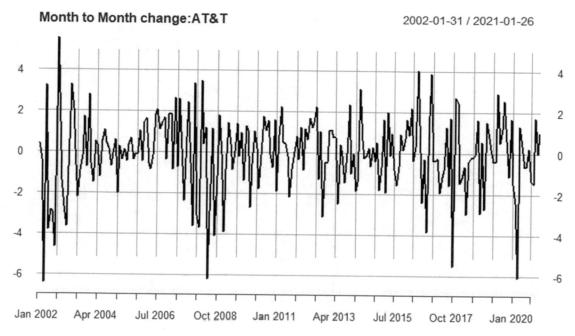

Figure 18-5. *Month over month variation for AT&T stock price, graphing with quantmod*

CHAPTER 19

Validation of Data

After spending five minutes doing data science, everyone knows that data preparation, including validation, is the most time-consuming step of any analysis. Several cleanup packages have been developed, including janitor and validate. Figure 19-1, from the validate package, shows a convenient graphic of three mtcars variables. It meets the data science trifecta: simple, quick, and handy.

In this code, the horsepower criteria of "more than 250" becomes validation rule V1. Most cars from this dataset fail that test, so the V1 graphic is nearly all red. Same logic applies to V2 and V3. It is a great way to look at many variables in light of some pass/fail or yes/no criteria.

```
if (!require("validate")) install.packages("validate")
big.engines <- check_that(mtcars, hp >250, cyl >5,  disp < 300)
big.engines
summary(big.engines)
```

```
##    name items passes fails nNA error warning expression
## 1    V1    32      2    30   0 FALSE   FALSE    hp > 250
## 2    V2    32     21    11   0 FALSE   FALSE     cyl > 5
## 3    V3    32     21    11   0 FALSE   FALSE disp < 300
```

```
plot(big.engines)
```

© William Yarberry 2021
W. Yarberry, *CRAN Recipes*, https://doi.org/10.1007/978-1-4842-6876-6_19

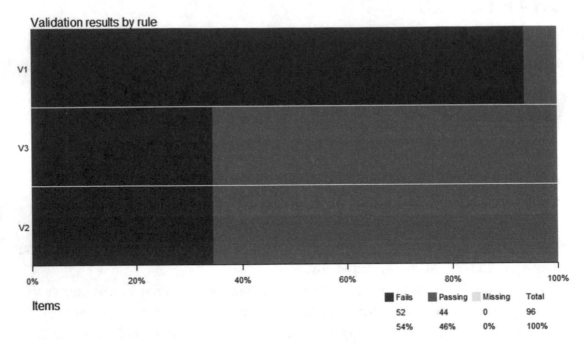

Figure 19-1. *View of three mtcars variables in terms of "pass" or "fail"*

This next example, adapted from the validate CRAN manual,[1] checks to ensure that the years listed are four digits. The first year is only three digits and hence fails (FALSE).

```
df <- data.frame(id = 11001:11003, year = c("218","2019","2020"),
value = 1:3)
rule <- validator(field_length(year, 4), field_length(id, 5))
out <- confront(df, rule)
out <- as.data.frame(out)
out

##    name value          expression
## 1    V1 FALSE field_length(year, 4)
## 2    V1  TRUE field_length(year, 4)
## 3    V1  TRUE field_length(year, 4)
## 4    V2  TRUE   field_length(id, 5)
## 5    V2  TRUE   field_length(id, 5)
## 6    V2  TRUE   field_length(id, 5)
```

---

[1]https://cran.r-project.org/web/packages/validate/validate.pdf, p. 37, accessed on January 29, 2021.

## CHAPTER 20

Shortcuts and Miscellaneous

20.1 RStudio

You can certainly use notepad or base R to write your scripts, in the same sense as a dog could walk on his hind legs to get places. It works but not well. RStudio is built for the purpose of efficient, effective, and beginner-friendly script coding.

Some simple shortcuts:

- Ctrl-Shift-R creates a section label, which is nothing more than a consistently formatted comment. Unless you enter a large number of words, the length of the comment is always the same. If you use it consistently in your code, it serves as a familiar road sign for new chunks of code. It has not execution functionality but does improve code readability. Here's an example:

  ```
  # I typed this section label using Ctrl-Shift-R ----------
  ```

- Use View(x) to see a structured (Excel-like) view. Although you cannot change data with View, you can easily find, filter, and sort columns as needed. Figure 20-1 shows an example View with filter of "h" for the tool column.

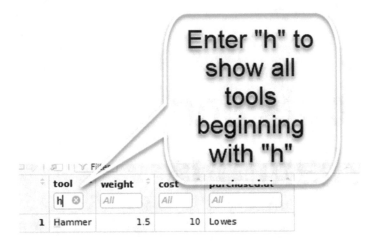

Figure 20-1. *Filtering in RStudio*

- Create a function from existing code. Highlight the code and press Ctrl-Alt-X. After you have done the conversion, be sure to run the function (highlight the function and press Ctrl-Enter) so that it is in memory, ready to be used. Figure 20-2 shows the process to create a simple function to double the value of a number.

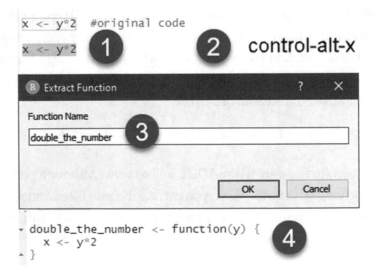

Figure 20-2. *Creation of a new function based on existing code using the Ctrl-Alt-X shortcut*

- Access many dozens of helpful debugging and navigation features using the shortcuts. Go to help and then keyboard shortcuts help (Alt-Shift-K). Figure 20-3 is a screenshot of the shortcuts screen after typing the shortcut. This display is semitransparent and disappears by pressing the Escape key. In the upper right-hand corner, click "See All Shortcuts" to get an easier-to-read list of all of them.

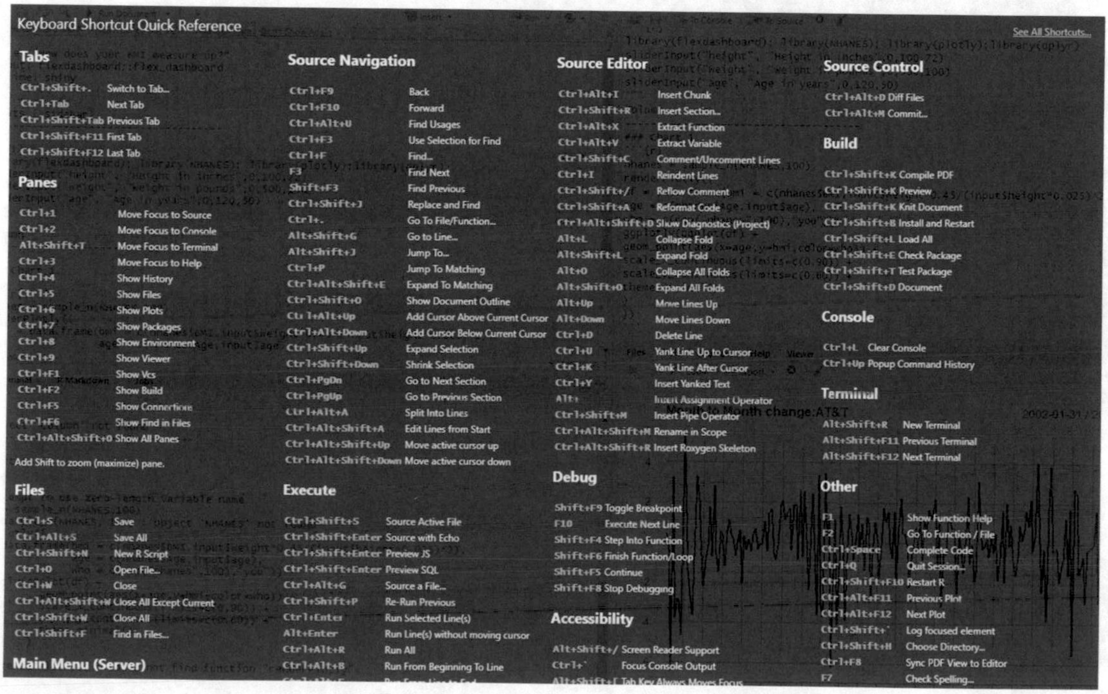

Figure 20-3. *List of most RStudio shortcuts using Alt-Shift-K*

- Create a Microsoft Word, PDF, or html document showing code, runtime messages, and results entirely in one document. R Markdown and the knitr package are beyond the scope of this book (although they are cool!), but you can create an attractive, reproducible document by simply pressing Ctrl-Shift-K. Figure 20-4 shows the process for a simple two-line script.

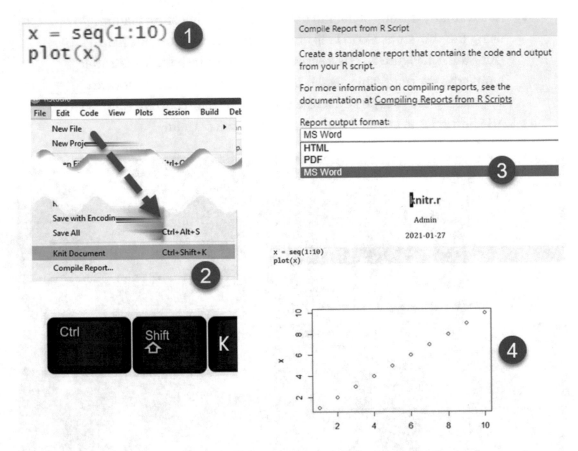

Figure 20-4.   *Create an MS Word document using knitr. Includes code, runtime messages, and output*

Create a data.frame on the fly for testing:

```
test.df <- data.frame( tool = c("Hammer", "Screwdriver","Saw"),
                       weight = c(1.5,0.5,1.2), cost = c(10,5, 12),
                       purchased.at = c("Lowes","ACE", "Home Depot"))
```

20.2 Utilities

20.2.1 Remove Read-Only Property from a Windows File

```
setwd("h:\\t")   #your working directory setting will vary
system("attrib -r test.file.docx")
```

```
## [1] 0
# return 0 means the action was successful
```

20.2.2 Rename a Windows File

assumes that my.word.docx exists in working directory; note random # added to renamed file name:

```
from.file <- c("my.word.docx")
another.name <- sample(seq(1:99999),size = 1)
another.name <- paste("new.word.doc.name",as.character(another.name), ".docx")
to.file <- another.name; file.rename(from.file, to.file);to.file
```

```
## Warning in file.rename(from.file, to.file): cannot rename file 'my.word.docx'
## to 'new.word.doc.name 71317 .docx', reason 'The system cannot find the file
## specified'
## [1] FALSE
## [1] "new.word.doc.name 71317 .docx"
```

20.3 Debugging R Code

R's error messages have occasionally been given some glum names, such as obscure, obtuse, obnoxious, and even odiferous. I will not be disrespectful here, but I can understand the feeling when the hours go by without resolution.

Fortunately, there's progress. tidyverse packages and some others will ask meaningful questions such as "did you mean …." and often suggest something helpful. R is open source and has developed over time with many contributors and many packages. As a result, interoperability challenges between its components have contributed to many of the nonobvious error messages. For example, as mentioned earlier in the book, if you load the package Zeilig after tidyverse, the select command will be the Zeilig's select version, not tidyverse's version. You'll get an error message that will frustrate you to no end because you probably coded the select command correctly—as required by tidyverse/DPLYR but not by Zeilig.

Here's a few general coding errors you'll likely run into when developing R scripts:

- Use of a single quote mark (') instead of a double quote mark (").

- Leaving a space between the assignment operator's characters "< -" rather than the correct "<-".

- Getting the class wrong for a variable. For example, your date column may be character rather than POSIX.

- Something you don't want is left over in memory from a previous session. To ensure that your program starts fresh, enter the command rm(list = ls()).

- Using a categorical variable where a numeric one is required or vice versa.

- Improper handling of missing data (NAs).

RStudio has three debugging tools:

- RStudio's error inspector and traceback() which list the sequence of calls that lead to the error

- RStudio's "Rerun with Debug" tool and options(error = browser) which open an interactive session where the error occurred

- RStudio's breakpoints and browser() which open an interactive session at an arbitrary location in the code

If you use these tools, combined with code using meaningful names and a clear structure, you should be able to get past most errors. If your data is voluminous, avoid the temptation to use your entire data to develop code. Use a subset first. Also, if you have code which looks for a rare condition, test by deliberately inserting a "hit" so that you know it works. Otherwise, you don't know whether you have bad code or your data genuinely has no records matching the condition.

Another technique I've used often is to do a cut and paste from the RStudio log error message to Google search (copy exactly; do not paraphrase). Remarkably, even error messages from lesser-known packages are discussed in forums. Most of the time you'll find your error or at least a workaround.

CHAPTER 21

Conclusion

My intent in writing this book was to provide useful code that you can put to work quickly. Some of the topics, such as the prediction models, are the subject of thick books and thousands of mathematical, scientific, and business-related papers. I wanted to show that with a few lines of code *some* functionality is possible without investing months or years going up the learning curve. Decades ago, those who controlled the old mainframes were called "high priests"—not necessarily a complimentary term. R has the potential to help people across the planet get value from data and so should be democratized as much as possible. Obviously if you have a need to do more in-depth analysis, there are terrific books available from Apress and others to get the expertise you need. But everyone has a "day one" in any subject. We should not discourage new users of the language by making it onerous to perform basic analysis. Success motivates.

Besides those relatively new to R, I had another group in mind as I wrote the book: those who use R intermittently and may forget specifics. What is outlined in this book is my best shot at R's frequently asked coding questions. For those who don't write books or haven't yet written one, here's a little secret: sometimes authors write books for themselves and just hope that others will benefit. As I run into needs in my business, I'll keep this book handy for reference. Sure, I could use Google, but then I might have to wade through a lot of irrelevant material. Like Sergeant Friday in the TV detective series said, "...just the facts, ma'am."

I'd love to hear from you. Suggestions, corrections, and even useful rants are welcome.

William Yarberry, CPA
Houston, Texas
January 31, 2021
byarberry@iccmconsulting.net

© William Yarberry 2021
W. Yarberry, *CRAN Recipes*, https://doi.org/10.1007/978-1-4842-6876-6_21

APPENDIX A

Suggested Websites

General RegEx information

 `www.regular-expressions.info/rlanguage.html`

 For extended regular expressions

 `www.boost.org/doc/libs/1_71_0/libs/regex/doc/html/boost_regex/syntax/`
`basic_extended.html`

 Cleaning data

 `https://cran.r-project.org/web/packages/textclean/readme/README.`
`html#html`

 `www.madebyspeak.com/blog/posts/become-a-master-text-wrangler-with-`
`regular-expressions`

 Whitespaces

 `https://regexone.com/lesson/whitespaces`

 Extracting web links

 `https://stackoverflow.com/questions/15579119/extract-websites-links-`
`from-a-text-in-r.`

 Locale documentation

 `www.rdocumentation.org/packages/readr/versions/1.3.1/topics/locale.`

 Word boundary

 `https://stackoverflow.com/questions/1324676/what-is-a-word-boundary-in-`
`regex`

 Negation

 `https://stackoverflow.com/questions/46898699/r-regex-for-containing-`
`pattern-with-negation`

 Common operators

 `http://web.mit.edu/gnu/doc/html/regex_3.html`

© William Yarberry 2021
W. Yarberry, *CRAN Recipes*, https://doi.org/10.1007/978-1-4842-6876-6

RegEx in Notepad++

http://blog.hakzone.info/posts-and-articles/editors/understanding-regex-with-notepad/comment-page-1/#comments

Wrappers for common string operations

www.rdocumentation.org/packages/stringr/versions/1.4.0

RStudio cheat sheet

https://rstudio.com/wp-content/uploads/2016/09/RegExCheatsheet.pdf

General discussion of regular expressions in R

www.gastonsanchez.com/r4strings/regex1.html

Suggestion for data science questions

https://towardsdatascience.com/20-questions-to-ask-prior-to-starting-data-analysis-6ec11d6a504b.

Cheat Sheet for RegEx in R

See also the nicely formatted cheat sheet from RStudio:

https://rstudio.com/wp-content/uploads/2016/09/RegExCheatsheet.pdf

RegEx entry	Meaning
\\d	Digits, 0, 1, 2...9.
\\D	Not digit.
\\s	Space.
\\S	Not space.
\\w	Word.
\\W	Not word.
\\t	Tab.
\\n	Newline.
^	Beginning of the string.
$	End of the string.
\	Escape special characters, for example, \\ is "\", \+ is "+".
		Alternation match, a.k.a. "OR."
•	Any character, except \n or line terminator.
[ab]	a or b.
[^ab]	Any character except a and b.

(continued)

W. Yarberry, *CRAN Recipes*, https://doi.org/10.1007/978-1-4842-6876-6

RegEx entry	Meaning
[0-9]	All digits.
[A-Z]	All uppercase A–Z letters.
[a-z]	All lowercase a–z letters.
[A-z]	All uppercase and lowercase a–z letters.
i+	i at least one time.
i*	i zero or more times.
i?	i zero or one time.
i{n}	i occurs n times in sequence.
i{n1,n2}	i occurs n1– n2 times in sequence.
i{n1,n2}?	Non-greedy match; see the preceding example.
i{n,}	i occurs >= n times.
[:alnum:]	Alphanumeric characters: [:alpha:] and [:digit:].
[:alpha:]	Alphabetic characters: [:lower:] and [:upper:].
[:blank:]	Blank characters, for example, space and tab.
[:cntrl:]	Control characters.
[:digit:]	Digits: 0 1 2 3 4 5 6 7 8 9.
[:graph:]	Graphical characters: [:alnum:] and [:punct:].
[:lower:]	Lowercase letters in the current locale.
[:print:]	Printable characters: [:alnum:], [:punct:], and space.
[:punct:]	Punctuation characters: ! " # $ % & ' ( ) * + , - . / : ; < = > ? @ [\ ] ^ _ ` {	} ~
[:space:]	Space characters: tab, newline, vertical tab, form feed, carriage return, and space.
[:upper:]	Uppercase letters in the current locale.
[:xdigit:]	Hexadecimal digits: 0 1 2 3 4 5 6 7 8 9 A B C D E F a b c d e f.

APPENDIX C

General R Comments by John D. Cook, Consultant[1]

John's broad-level comments on regular expressions in R are useful as a level set. The perspective of someone using different flavors of RegEx gives us context. Here's a quote from one of his web pages:

> The R language was developed for analyzing data sets, not for munging text files. However, R does have some facilities for working with text using regular expressions. This comes in handy, for example, when selecting rows of a data set according to regular expression pattern matches in some columns.
>
> R supports two regular expression flavors: POSIX 1003.2 and Perl. Regular expression functions in R contain two arguments: extended, which defaults to TRUE, and perl, which defaults to FALSE. By default R uses POSIX extended regular expressions, though if extended is set to FALSE, it will use basic POSIX regular expressions. If perl is set to TRUE, R will use the Perl 5 flavor of regular expressions as implemented in the PCRE library.
>
> Regular expressions are represented as strings. Meta-characters often need to be escaped. For example, the metacharacter \w must be entered as \\w to prevent R from interpreting the leading backslash before sending the string to the regular expression parser.
>
> The grep function requires two arguments. The first is a string containing a regular expression. The second is a vector of strings to search for matches. The grep function returns a list of indices. If the regular expression matches a particular vector component, that component's index is part of the list.

---

[1]"Regular Expressions in R," accessed on March 27, 2020, www.johndcook.com/blog/r_language_regex/.

W. Yarberry, *CRAN Recipes*, https://doi.org/10.1007/978-1-4842-6876-6

Example:

grep("apple", c("crab apple", "Apple jack", "apple sauce")) returns the vector (1, 3) because the first and third elements of the array contain "apple." Note that grep is case-sensitive by default and so "apple" does not match "Apple." To perform a case-insensitive match, add ignore.case = TRUE to the function call.

There is an optional argument value that defaults to FALSE. If this argument is set to TRUE, grep will return the actual matches rather than their indices.

The function sub replaces one pattern with another. It requires three arguments: a regular expression, a replacement pattern, and a vector of strings to process. It is analogous to s/// in Perl. Note that if you use the Perl regular expression flavor by adding perl = TRUE and want to use capture references such as \1 or \2 in the replacement pattern, these must be entered as \\1 or \\2.

The sub function replaces only the first instance of a regular expression. To replace all instances of a pattern, use gsub. The gsub function is analogous to s///g in Perl.

The function regexpr requires two arguments, a regular expression and a vector of text to process. It is similar to grep, but returns the locations of the regular expression matches. If a particular component does not match the regular expression, the return vector contains a -1 for that component. The function gregexpr is a variation on regexpr that returns the number of matches in each component.

The function strsplit also uses regular expressions, splitting its input according to a specified regular expression.

For comments or questions, John D. Cook can be reached at cook@johndcook.com.

APPENDIX D

Understanding a Long Regular Expression

The following is courtesy of the Turing School of Software and Design.[1]

Understanding a Long Regular Expression

/^((([a-zA-Z]|[0-9])|([-]|[_]|[.]))+[@]((([a-zA-Z0-9])|([-])){2,63}[.](([a-zA-Z0-9]){2,63})+$/gi

The preceding RegEx is a pattern for a simple email matcher. If we copy and paste this RegEx into a site like regex101, we can start to try to figure out what's going on.

The Base Layer

Let's look specifically at the setup for this RegEx pattern.

/^$/gi

Character	Description
/	Indicates the beginning of a RegEx pattern
^	Anchor: Indicates that the match must be at the beginning of the string
$	Anchor: Indicates that the match must reach to the end of the string
/	Indicates the end of a RegEx pattern

(continued)

---

[1]"RegEx Fun Times - Front-End Engineering Curriculum - Turing School of Software and Design," accessed March 23, 2020, http://frontend.turing.io/.

© William Yarberry 2021

W. Yarberry, *CRAN Recipes*, https://doi.org/10.1007/978-1-4842-6876-6

Character	Description
g	Flag: Outside the RegEx meaning "global." That means it will test the pattern against all possible matches in a string. Without this flag, the RegEx will only test the first match it finds and then stop.
I	Flag: Match is case insensitive.

Capturing a Character

([a-zA-Z]|[0-9])

Character	Description
( )	Specifies the beginning of a grouping. For example, dog|dig and d(o|i)g can both be used to capture dog or dig strings.
[ ]	Creates a character class
-	Is a range
G	Flag: Outside the RegEx meaning "global." That means it will test the pattern against all possible matches in a string. Without this flag, the RegEx will only test the first match it finds and then stop.
a-z	Is any character in the range between a through z
[a-zA-Z]	So put together, this character class would match any letter a through z, case insensitive.
[0-9]	A character group that would match any character in a range between 0 and 9
I:	Is a Boolean "or"

So all put together, ([a-zA-Z]|[0-9]) matches any letter of number.

Matching multiple times:

(([a-zA-Z]|[0-9])|([-]|[_]|[.]))+

We can now see that this RegEx pattern nests a capture group within another capture group to find any letter or number or a -, _, or .

The interesting piece here is the character at the very end: +.

This is a quantifier which targets the capture group and says "repeat previous token 1 to infinite times" in your definition of a match.

Quantifiers

Character	Description
?	May include zero or one occurrence of the preceding element
*	May include zero, one, or many occurrences of the preceding element
+	Must include one or many occurrences of the preceding element
{n}	Must match the preceding element n times
{min,}	Must match the preceding element at least min or more times
{min,max}	Must match the preceding element at least min times but not more than max times
{,max}	Must match the preceding element not more than max times

Final Breakdown

/^((([a-zA-Z]|[0-9])|([-]|[_]|[.]))+[@]((([a-zA-Z0-9])|([-])){2,63}[.]((([a-zA-Z0-9]){2,63})+$/gi

/^ – Start a regular expression at the beginning of the string.

((([a-zA-Z]|[0-9])|([-]|[_]|[.]))+ – Match any letter, number, or -, _, . one or more times until...

[@] – Match an @ symbol.

((([a-zA-Z0-9])|([-])){2,63} – Match any letter, number, or - 2 to 63 times... why 63?

[.]– Match a period.

((([a-zA-Z0-9]){2,63})+ – Match any letter or number 2 to 63 times, one or many times.

$/ – Anchor this match to the end of the string.

gi – And make it a global and case-insensitive match.

Regular Expression Enabled Languages

The following is kindly provided by RegExBuddy.[1]

If you are a programmer, you can save a lot of coding time by using regular expressions. With a regular expression, you can do powerful string parsing in only a handful lines of code or maybe even just a single line. A RegEx is faster to write and easier to debug and maintain than dozens or hundreds of lines of code to achieve the same by hand.

- Boost – Free C++ source libraries with comprehensive RegEx support that was later standardized by C++11. But there are significant differences in Boost's RegEx flavors and the flavors in std::regex implementations.

- Delphi – Delphi XE and later ship with regular expressions and regular expression core units that wrap the PCRE library. For older Delphi versions, you can use the TPerlRegEx component, which is the unit that the regular expression core unit is based on.

- Gnulib – Gnulib or the GNU portability library includes many modules, including a RegEx module. It implements both POSIX flavors, as well as these two flavors with added GNU extensions.

---

[1]"Popular Tools, Utilities and Programming Languages That Support Regular Expressions," accessed on April 8, 2020, www.regular-expressions.info/tools.html.

W. Yarberry, *CRAN Recipes*, https://doi.org/10.1007/978-1-4842-6876-6

- Groovy – Groovy uses Java's java.util.regex package for regular expression support. Groovy adds only a few language enhancements that allow you to instantiate the Pattern and Matcher classes with far fewer keystrokes.

- Java – Java 4 and later include an excellent regular expression library in the java.util.regex package.

- JavaScript – If you use JavaScript to validate user input on a web page at the client side, using JavaScript's built-in regular expression support will greatly reduce the amount of code you need to write.

- .NET (dot net) – Microsoft's new development framework includes a poorly documented but very powerful regular expression package that you can use in any .NET-based programming language such as C# (C sharp) or VB.NET.

- PCRE – Popular open source regular expression library written in ANSI C that you can link directly into your C and C++ applications or use through a .so (UNIX/Linux) or a .dll (Windows).

- Perl – The text-processing language that gave regular expressions a second life and introduced many new features. Regular expressions are an essential part of Perl.

- PHP – Popular language for creating dynamic web pages, with three sets of RegEx functions. The first two implement POSIX ERE, while the third is based on PCRE.

- POSIX – The POSIX standard defines two regular expression flavors that are implemented in many applications, programming languages, and systems.

- PowerShell – Windows PowerShell is a programming language from Microsoft that is primarily designed for system administration. Since PowerShell is built on top of .NET, its built-in RegEx operators -match and -replace use the .NET RegEx flavor. PowerShell can also access the .NET RegEx classes directly.

- Python – Popular high-level scripting language with a comprehensive built-in regular expression library.

- R – The R language is the programming language used in the R project for statistical computing. It has built-in support for regular expressions based on POSIX and PCRE.

- Ruby – Another popular high-level scripting language with comprehensive regular expression support as a language feature.

- std::regex – RegEx support part of the standard C++ library defined in C++11 and previously in TR1.

- Tcl – Tcl, a popular "glue" language, offers three RegEx flavors: two POSIX-compatible flavors and an "advanced" Perl-style flavor.

- VBScript – Microsoft scripting language used in ASP (Active Server Pages) and Windows scripting, with a built-in RegExp object implementing the RegEx flavor defined in the JavaScript standard.

- Visual Basic 6 – Last version of Visual Basic for Win32 development. You can use the VBScript RegExp object in your VB6 applications.

- wxWidgets – Popular open source windowing toolkit. The wxRegEx class encapsulates the "Advanced Regular Expression" engine originally developed for Tcl.

- XML Schema – The W3C XML Schema standard defines its own regular expression flavor for validating simple types using pattern facets.

- Xojo – Cross-platform development tool formerly known as REALbasic, with a built-in RegEx class based on PCRE.

- XQuery and XPath – The W3C standard for XQuery 1.0 and XPath 2.0 Functions and Operators extends the XML Schema RegEx flavor to make it suitable for full-text search.

- XRegExp – Open source JavaScript library that enhances the RegEx syntax and eliminates many cross-browser inconsistencies and bugs.

Sample Data Analysis Questions

Successful data scientists or even casual users of analytics tools should keep a yellow sticky note with the following words pasted to the bottom of their monitors:

Have I asked the right questions?

The following is a list of questions you might potentially ask at the beginning of an analytics project. Of course, it will vary by business/scientific domain, but many of these questions apply across the board.

Numerics:

- Largest numbers

- Smallest numbers

- Duplicate numbers

- Averages

- Modes (most frequent numbers)

- Medians (middle numbers)

- Standard deviations

- Coefficients of variation

- Gaps in the data

- Number of records

- Number of records with bad data, such as alphas in numeric fields

© William Yarberry 2021
W. Yarberry, *CRAN Recipes*, https://doi.org/10.1007/978-1-4842-6876-6

- Range of values

- Is the numeric data normally distributed? For example, in R if you run the shapiro.test against highly non-normal data, such as 1,1,1,1,1, 1,1,1,1,1,1,1,1,1,2,2,2,2,2,2,2,2,2, then a very small P value (9e-07) will result.

- What is the skewness of the sample or population? Is it zero, negative, or positive?

- What is the kurtosis of the sample or population?

- What is the distribution of the sample or population? Have you identified the distributions of your key numeric predictors? Some distribution families include beta, binomial, chi-squared, exponential, F, gamma, geometric, Weibull, hyper-geometric, log-normal, logistic, multinomial, negative binomial, normal, Poisson, Student's t, Bernoulli, and uniform distributions.

- What are the measures of spread? These would include minimum, first quartile, median, third quartile, maximum, range, and interquartile range. The Tukey five-number test is often used for a quick look at the data.

- What do common values look like? Randomly sample the data, particularly when there are hundreds of thousands or more records. This gives you a sense of what's in the files without tediously scrolling through many rows.

- Does anything pop out when you convert all numbers to absolute values?

- Run a histogram and density plot.

- Run a simple line graph of numeric values. Are there any trends visible?

- If the data is a time series, what patterns are shown in the data? Effects such as seasonal movements, cyclical movements, irregular fluctuations, and lagging (among other factors) should be considered. Single spectrum (Fourier) analysis is also used to model time series.

Categorical (character) data:

- Has a Chi-squared test for categorical data been performed? Suppose there is a city of 1,000,000 residents with four neighborhoods: A, B, C, and D. A random sample of 650 residents of the city is taken, and their occupation is recorded as "white collar," "blue collar," or "no collar." The null hypothesis is that each person's neighborhood of residence is independent of the person's occupational classification. After performing the calculations, a very large test statistic rejects the null hypothesis of independence.[1] The actual case is that neighborhood of residence is affected by the occupational classification.

- What is the median of ordinal categorical data? Of course, the answer will be categorical.

More detailed questions/considerations:

- Do ties (duplicates or near duplicates) exist in the data? If so, the data could potentially have a non-normal distribution.

- Do you or your team have enough domain knowledge to separate correlation and causation?

- Are random forests an appropriate technique for your data prediction model? If you have large datasets, the random forest technique may be preferred due to higher computer efficiency.

- Is your data structured or unstructured? If unstructured, what tools or approach will you use to format and clean the data?

- What questions will your audience ask? For example, will they ask for filtering on keys, structures, categories, or ordinate numeric values? Will they be looking for geographic/mapping of data or dynamic trends over time? And will there be an ability to drill down into details during presentations to colleagues and stakeholders?

---

[1] "Chi-Squared Test," Wikipedia, accessed on July 21, 2019, https://en.wikipedia.org/wiki/Chi-squared_test.

- What normalization will be required so that different groups or sets with varying scales can be compared?

- Have you considered removing outliers with a trim argument to remove the highest and lowest observations?

- For correlation, will you use the Pearson statistic, which assumes normal distribution of the data; the Spearman correlation, which is nonparametric (assumes nothing about the underlying distribution); or Kendall's rank correlation coefficient, which is based on ranking of the values rather than the values themselves?

- When looking at observations in the millions or billions, have you used visual tools such as scatterplots to facilitate your understanding of the data characteristics?

- Have you identified or created model variables which are not correlated with each other by using principal component analysis (PCA)?

- Will bootstrap resampling mitigate the effect of too many or too large outliers? These outliers can significantly distort your prediction model.

- Have categorical predictors been summarised in a table, showing odds and odds ratios?

- For numeric fields, what are the totals for net, positive, negative, and absolute value?

- Where appropriate, has numeric data been standardized using transformations such as the Z score?

- Do duplicate numbers exist? Are there duplicate character values where only one entry per table would be expected? For example, a contest may be structured so that there should be only one top contestant and the same contestant should not be listed more than once.

- Are there missing sequence numbers?

- Are there repeated, specific numbers, such as "40,234.1"? How likely is that to be legitimate based on what you know about the data? What does a numeric frequency analysis look like? Does the population or sample mode look reasonable?

- Would it make sense to run Benford tests on the data? Many "natural" data, such as populations of cities and some tax information, have recognizable differences in frequencies of first, second, and sometimes third digits. Deviations from those patterns may suggest, but not prove, purposeful manipulation of the data (unauthorized or unintentional).

- Does an odd size relationship exist between the highest and second highest number? Lowest and second lowest number?

- Do gaps exist and are they consistent? Are some gaps large relative to the other gaps? This includes missing data, such as weather data, where an observation is expected every hour or day. Significant gaps in numeric data, such as the absence of numbers starting with a particular digit, should be investigated.

- Are numeric fields numeric?

- Are character fields uniformly shown as string data? Is case consistent? Does it need to be consistent for your analysis to work?

- Is data sorted in the expected sequence? Does it need to be?

- If various types of joins are needed, does the matching data have referential integrity? For example, if an analysis of chemical compositions uses a standard chemical batch number, will another table containing supplementary batch information have the same number? Will there be mismatches?

- If you are working with large datasets on your organization's servers, have you double- and triple-checked SQL extractions for logic flaws? For example, using an outer join and a predicate incorrectly placed could result in unreasonable runtimes.

- Are unit measures consistent? For example, will part of the data be in kilograms and the rest in pounds? According to *Times Science*, in 1999, "NASA lost its $125-million Mars Climate Orbiter because spacecraft engineers failed to convert from English to metric measurements when exchanging vital data before the craft was launched...."

- Does the data have obvious outliers which can be clearly identified as errors and eliminated from the dataset?

- Does the data potentially have outliers which require additional analysis because they are not sufficiently different from expected data?

- Are any expected data points missing? For example, are you expecting data from each nation in the European Union, but France is missing?

- Do standard measures of variation show unreasonable values? Perhaps you are working with five years of pricing data for a part. Would a coefficient of variation of 50% make sense?

- For time series data, are cutoffs for each period (week, month, quarter, etc.) accurate?

- If monthly or yearly data is being compared, are differences in number of days considered? If extreme precision is required, are periodic time adjustments taken into consideration? For example, "leap minutes" due to wobbles in the earth's orbit may need to be considered.

Visualizations checklist:

Will any of the following visualizations facilitate your understanding of the data or help present your findings?

- Bar charts

- Stacked bar charts

- Stacked area charts

- Dual-axis combination chart (multiple bar graphs with line graphs, all on same page)

- Sparklines

- Pie charts

- Density plots

- Scatterplots

- Line graphs

- Line graphs with smoothing, regression lines and other embedded analytics

- Histograms

- Frequency polygons

- Time series

- Area plots, including geocoding

- Text plots

- Dot plots

- Plots with vertical lines connecting to points

- Correlograms

- Box and whisker plots

- Surface plots

- Mapping plots

- Choropleth maps

- Hex plots

- Faceting plots (multiple plots on the same page)

- Violin plots

- Gradient color plots

- Graphs based on polar coordinates

- Step charts

- Correlation graphs

- Venn diagram

- Survival curves

- Dynamic trend analyzer, which graphs changes over time. The late Hans Rosling popularized these techniques and developed the Trendalyzer system. Google acquired the software and has made parts of it available to the public. See TED Talks for some impressive demonstrations of the visual displays of demographic information.

- Word cloud to show higher-frequency words used in text

- Streamgraphs

- Treemaps

- Heatmaps

- Donut charts

- Interactive charts (online)

- Pairs/scatterplots to show graphical relationships between variables (are they correlated?)

- Bubble charts and moving bubble charts

- Histogram and density plots together

- Nightingale charts

- Parallel coordinates plots (using lattice package in R)

- Model-based clustering charts

- Location trace maps (also called sequential path maps)

- Animated growth maps

- Gantt charts

- Waterfall charts

- Slope graphs (single or dual axis)

- Funnel charts

- Pace charts. They are sometimes called bullet charts—a horizontal bar chart with each bar containing a narrow sub bar showing progress against the main bar.

- Pareto charts

- Bump charts (show how some categorical variable changes numerically over time)

- Dumbbell charts

- Dashboard-type visualizations with user-controlled, dynamic views

- One-dimensional unit charts

- Have you used a "dodge" function to increase the visibility of data points which may otherwise obscure each other? Both Python and R contain this function.

- Beyond graphical displays, have you considered tools to improve the appearance of tables of numbers? For example, the formattable package in R performs this function.

Formats Recognized by Lubridate

Source: Official CRAN manual for the Lubridate package, `https://cran.r-project.org/web/packages/lubridate/lubridate.pdf`.

Code	Description
a	a Abbreviated weekday name in the current locale (also matches full name).
A	A Full weekday name in the current locale (also matches abbreviated name). You don't need to specify a and A formats explicitly. Wday is automatically handled if preproc_wday = TRUE.
b	b (!) Abbreviated or full month name in the current locale. The C parser currently understands only English month names.
B	B (!) Same as b.
d	d (!) Day of the month as decimal number (01–31 or 0–31).
H	H (!) Hours as decimal number (00–24 or 0–24).
I	I (!) Hours as decimal number (01–12 or 1–12).
j	j Day of year as decimal number (001–366 or 1–366).
q	q (!*) Quarter (1–4). The quarter month is added to the parsed month if m format is present.
m	m (!*) Month as decimal number (01–12 or 1–12). For parse_date_time. As a Lubridate extension, also matches abbreviated and full months' names as b and B formats. C parser understands only English month names.

(continued)

© William Yarberry 2021
W. Yarberry, *CRAN Recipes*, https://doi.org/10.1007/978-1-4842-6876-6

Code	Description
M	M (!) Minute as decimal number (00–59 or 0–59).
p	p (!) AM/PM indicator in the locale. Normally used in conjunction with I and not with H. But the Lubridate C parser accepts H format as long as hour is not greater than 12. C parser understands only English locale AM/PM indicator.
S	S (!) Second as decimal number (00–61 or 0–61), allowing for up to two leap seconds (but POSIX-compliant implementations will ignore leap seconds).
OS	OS Fractional second.
U	U Week of the year as decimal number (00–53 or 0–53) using Sunday as the first day of the week (and typically with the first Sunday of the year as day 1 of week 1). The US convention.
w	w Weekday as decimal number (0–6, Sunday is 0).
W	W Week of the year as decimal number (00–53 or 0–53) using Monday as the first day of the week (and
ty	typically with the first Monday of the year as day 1 of week 1). The UK convention.
y	y (!*) Year without century (00–99 or 0–99). In parse_date_time(), also matches year with century (Y format).
Y	Y (!) Year with century.
z	z (!*) ISO8601 signed offset in hours and minutes from UTC. For example, –0800, –08:00, and –08 all represent 8 hours behind UTC. This format also matches the Z (Zulu) UTC indicator. Because base::strptime() doesn't fully support ISO8601, this format is implemented as a union of four orders: Ou (Z), Oz (–0800), OO (–08:00), and Oo (–08). You can use these four orders as any other, but it is rarely necessary. parse_date_time2( ) and fast_strptime( ) support all of the time zone formats.
Om	Om (!*) Matches numeric months and English alphabetic months (both long and abbreviated forms).
Op	Op (!*) Matches AM/PM English indicator.
r	r (*) Matches Ip and H orders.
R	R (*) Matches HM and IMp orders.
T	T (*) Matches IMSp, HMS, and HMOS orders.

Other Books by Bill Yarberry

$250K Consulting

- *501 Data Science Questions*
- *The Effective CIO*
- *What Top CIOs Know*
- *GDPR: A Short Primer*
- *Computer Telephony Integration*
- *Telecommunications Cost Management*
- *Write a Business Plan to Make Money*

© William Yarberry 2021
W. Yarberry, *CRAN Recipes*, https://doi.org/10.1007/978-1-4842-6876-6

Index

A

ASCII coding, 221

B

Built-in datasets, 1, 4, 15, 38, 40, 241, 242

C

Character classes
   anchor syntax, 185, 186
   dot, 183, 184
   extended regular expression, 189
   locale() set, 189, 190
   multiline specification, 186
   range, 183
   whitespace, 187
   word boundaries, 186
Coordinated Universal Time zone
   (UTC), 130

D

Data analysis tools
   categorical data, 327
   detailed questions/considerations,
     327–330
   monitors, 325
   numerics, 325, 326
   visualizations checklist, 330–333

Data structures, 241, 242, 327
Data validation, 301, 302
Date and time processing
   alignment, 118, 119
   calculate duration, 112–114
   calculations, 114–116
   date validation, 139
     automatic roll dates, 159, 160
     difference time zone, 148
     eastern daylight savings
      time, 149, 150
     internationalization, 151–153
     locale names, 152
     make_difftime() function, 144–146
     right now, 152, 153
     rollback, 153, 154
     rounding function, 155–159
     shorthand methods, 142
     test interval/date, 143, 144
     time difference, 139, 140
     time zone codes, 141, 142
     time zones, 146, 147
     week calculation, 142
   duration calculations, 110
   hard-coding coffee time, 109
   overlaps, 116, 117
   parse_date_time, 138, 139
   period and duration, 122, 123
   periods record, 119, 120
   sequencing, 120–122

Printed in the United States
by Baker & Taylor Publisher Services